Lecture Notes in Mathematics

Edited by A. Dold and B. Eckmann

Subseries: Mathematisches Institut der Universität und
Max-Planck-Institut für Mathematik, Bonn – vol. 3
Adviser: F. Hirzebruch

1101

Vincent Cossart
Jean Giraud
Ulrich Orbanz

Resolution of Surface Singularities

Three Lectures with an Appendix by H. Hironaka
Edited by U. Orbanz

Springer-Verlag
Berlin Heidelberg New York Tokyo 1984

Authors

Vincent Cossart
Université Pierre et Marie Curie (Paris VI), Mathématiques
4, Place Jussieu, 75005 Paris, France

Jean Giraud
Université de Paris-Sud, Centre d'Orsay, Mathématique
Bât. 425, 91405 Orsay Cédex, France

Ulrich Orbanz
Max-Planck-Institut für Mathematik
Gottfried-Claren-Str. 26, 5300 Bonn, Federal Republic of Germany

AMS Subject Classification (1980): 14E15, 14J17

ISBN 3-540-13904-4 Springer-Verlag Berlin Heidelberg New York Tokyo
ISBN 0-387-13904-4 Springer-Verlag New York Heidelberg Berlin Tokyo

Printing and binding: Beltz Offsetdruck, Hemsbach/Bergstr.
2146/3140-543210

RESOLUTION OF SURFACE SINGULARITIES - THREE LECTURES

Introduction

The problem of resolution of the singularities of an algebraic surface has a long history reaching back to the last century. After its solution (by R. Walker in 1935), it was quite reasonable to ask for desingularization of algebraic varieties of arbitrary dimensions. For the general problem the surface case and the different methods of its solution were of special importance, not only as a tool for higher dimension, but also as a testing ground for the general case. To quote D. Hilbert: "Vielleicht in den meisten Fällen, wo wir die Antwort auf eine Frage vergeblich suchen, liegt die Ursache des Mißlingens darin, daß wir einfachere noch unvollkommen erledigt haben. Es kommt dann alles darauf an, diese leichteren Probleme aufzufinden und ihre Lösungen mit möglichst vollkommenen Hilfsmitteln und durch verallgemeinerungsfähige Begriffe zu bewerkstelligen."

Meanwhile the general problem of desingularization in characteristic zero has been solved by Hironaka, but there are some points about his proof which are a challenge for further investigations. First of all, since his proof is very complicated, it is natural to look for simplifications. One such simplification for instance is the notion of idealistic exponents, introduced by Hironaka himself. Then, of course, the case of positive characteristics remains open (in dimension > 3). I want to stress another point which seems unsatisfactory: If Hironaka's proof is specialized for surfaces, there is still no substantial simplification of his procedure. This is one reason to look for different methods to desingularize surfaces, thereby hoping to find some more "natural" method for desingularization in the general case.

Another point of view is the classification of singularities by their resolution (as in the case of plane curves). Of course, for this purpose one needs a canonical procedure for resolution. In a vast generalization of the plane curve case, Abhyankar has developed a machinery that will ultimately lead to a canonical procedure for desingularization. But there again the mechanism is too complicated to allow an easy description in dimension 2.

Yet another aspect of resolution in dimension > 1 is the problem of globalizing some local (or even punctual) algorithm. The most complicated part of Hironaka's paper is devoted to this problem and also in Abhyankar's proof of resolution of surfaces in positive characteristics, globalizing presents a serious difficulty.

To summarize, a good proof in the surface case should combine the following features: It should be canonical, it should be easy in some sense, and globalization should be no problem.

The starting point for a new development in the resolution of surfaces, after Hironaka's proof in 1964, was Zariski's paper [6][1] in which the author applied his new theory of equisingularity to give a new proof for resolution of surfaces. By using suitable projections as for equisingularity, he introduced the notion of quasi-ordinary singularities. He also extracted numerical invariants to measure the improvement under (suitably chosen) monoidal transformations. The result was the first canonical proof for resolution of surfaces that did not use normalization.

Abhyankar and Hironaka followed these ideas along different lines. Using a detailed description of the characteristic polyhedron, Hironaka modified the definition of a quasi-ordinary singularity to give another proof for resolution of surfaces, including characteristic $p > 0$ (see appendix). Abhyankar made the distinction between "good" and "bad" points, corresponding to their behaviour along a sequence of monoidal transformations. This distinction is the result of a careful analysis of Zariski's method of quasi-ordinary singularities, with the intention to suppress all properties of singularities which are not directly related to the process of resolution. A special feature of Abhyankar's method is that a slight (but very subtle) modification gives the stronger result of "embedded resolution" (see first lecture for explanation). This result in turn seems to have inspired Zariski to use his quasi-ordinary singularities for a different proof of embedded resolution in 1978 [7].

Now we come to the content of the contributions of this volume, all of which have been presented in a seminar on numerical characters of singularities, held in Bonn 1979-1981. The first lecture is a completely

---

[1] see the references of the first lecture

self-contained presentation of Abhyankar's method for resolution of surfaces, based on a lecture of Abhyankar's given in Paris in May 1980. Abhyankar's proof meets all requirements stated above: It is canonical, the local method globalizes trivially , and the proof is easy and short We note that meanwhile there are some efforts to use this proof for the classification of some surface singularities.

The second lecture by Giraud introduces the various formulations of resolution and related problems. Then some different methods for resolution of surfaces are sketched, and finally an attempt is made to describe Zariski's ideas of 1967 clearly, without actually giving the proof.

The third lecture by Cossart contains a detailed comparison of Hironaka's and Abhyankar's generalizations of the notion of quasi-ordinary singularities. Abhyankar's method is translated into the language of characteristic polyhedra and quasi-ordinary points in the sense of Hironaka in order to describe more precisely the links and the differences between the methods of Abhyankar and Hironaka. Abhyankar's method is extended to positive characteristics, but only non-embedded resolution is treated in this lecture.

Hironaka's lectures on the resolution of excellent surfaces of any characteristic, given at Bowdoin College and written down by B.M. Bennett, have been distributed privately for many years. Since these notes are a basic reference for the subject, they are included here as an appendix in order to make them available for the mathematical community.

The three lectures differ considerably in style, due to the intentions and tastes of the authors. The first lecture uses the language of local algebra, and it assumes only a little background in this field. The second lecture uses the language of modern Algebraic Geometry. Since this lecture does not contain proofs, some familiarity with this language is sufficient for the reader. Cossart's contribution is self contained modulo the Bowdoin College lectures by Hironaka (see appendix) and a technical result in one of Cossart's papers.

I gratefully acknowledge the support by the Max-Planck-Institut für Mathematik (Bonn) during the final preparation of this volume.

U. Orbanz

CONTENTS

EMBEDDED RESOLUTION OF ALGEBRAIC SURFACES AFTER ABHYANKAR (Characteristic 0)

Ulrich Orbanz

This lecture is based on four seminar talks given by S.S. Abhyankar in Paris between May 5 and May 8, 1980. I had the opportunity to attend these lectures and to discuss the topic with S.S. Abhyankar. We agreed that it would be worth having a written exposition of his talks, not only containing all the details that could not be given orally, but also making the proof accessible to non-specialists, assuming only some basic facts of commutative algebra.

The material presented by Abhyankar in Paris is the content of chapter III, and I tried to keep this close to his instructive oral exposition. My own contribution restricts itself to the arrangement of the background material given in chapters I, II and IV.

According to Abhyankar, his results contained in this lecture have been obtained by him in 1967, but the only publication on this proof was Lipman's sketch in his Arcata Lectures [3]. Meanwhile Abhyankar himself has written two papers on the subject, for which my lecture may serve as an introduction:

1) Desingularization of plane curves, in Singularities, Amer. Math. Soc., Proc. Symp. Pure Math. 40 (1983), 1-45.

This paper contains the results parallel to chapter II of the present lecture, given in the terminology developed in [8].

2) Good points of a hypersurface, to appear in the Advances in Mathematics.

As the title indicates, this long paper contains a generalization of the notion of a "good point" to any hypersurface, and the techniques for this notion are developed for any dimension. At the end, the embedded resolution of surfaces appears as an application of the more general theory.

I am very much indepted to Prof. Abhyankar for his continual support during the preparation of these notes.

Table of Contents

I Elementary properties of blowing up

§1 Conventions and preliminaries

All rings considered here are supposed to be commutative and to contain a unit element. By a local ring we mean a noetherian ring having a unique maximal ideal. If R is a local ring, the maximal ideal of R will be denoted by M(R) (and the definition of maximal ideal includes $M(R) \neq R$, so that $R \neq 0$). The completion of a local ring R will be denoted by R*.

For any local ring R, the function $\operatorname{ord}(R): R \to \mathbb{Z} \cup \{\infty\}$ is defined by

$$\operatorname{ord}(R)(f) = \sup \{n \in \mathbb{Z} \mid f \in M(R)^n\} , \; f \in R.$$

So by Krull's intersection theorem

$$\operatorname{ord}(R)(f) = \infty \Leftrightarrow f = 0.$$

For the theory of regular local rings we refer to [4] or [5]. We will make use of the facts that regular local rings are Unique Factorization Domains, and that any localization of a regular ring is regular again.

Furthermore we need some facts about excellent rings that can be found in [4]. If R is an excellent ring, then so is any localization of R and any finitely generated R-algebra (this includes homomorphic images). If R is an excellent domain, then the integral closure $\bar{R}$ of R in its quotient field is a finitely generated R-module. We will also use that the formal fibres of an excellent local ring R are regular. This means that given any prime ideal P of R and any prime ideal Q of R* such that $Q \cap R = P$, the local ring $(R^*)_Q/P(R^*)_Q$ is regular. In particular if R is reduced, then so is R*.

Next we define the notion of normal crossing for any set of ideals in a regular local ring.

<u>Definition.</u> Let R be a regular local ring and let I* be a set of ideals of R. I* is said to have normal crossings if R has a regular system of parameters $\underline{x} = \{x_1, \ldots., x_r\}$ with the following property: If P is any prime ideal of R associated to an ideal I of I*, then P is generated by a subset of $\underline{x}$.

Sometimes also elements of R will be said to have normal crossing, by which we mean that the principal ideals generated by these elements have normal crossings. For example, a nonzero element f of R has normal crossings if and only if there is a regular system of parameters $x_1, \ldots, x_r$ of R such that $f \cdot R$ is generated by a monomial in $x_1, \ldots, x_r$. There is an obvious way to translate the definition of normal crossing into geometry, i.e. to a regular point P on a noetherian scheme X and a set of closed subschemes passing through P.

§2 Blowing up

We will describe blowing up only locally. The verification of the global nature of this process is left to the reader.

Given a local ring R and an ideal I of R, the blowing up of R with center I is a family of homomorphisms $R \to R_1$ of local rings obtained in the following manner. Given any $x \in I$, $x \neq 0$, we consider the ring R_x obtained by inverting the powers of x. The subring of R_x generated by the image of R (under the canonical homomorphism) and the elements a/x, $a \in I$, will be denoted by $R[I/x]$. (Note that $R[I/x]$ is generated over R by $a_1/x, \ldots, a_n/x$, where $a_1, \ldots, a_n$ is any system of generators of I. Therefore $R[I/x]$ is noetherian.) The homomorphism $R \to R_1$ belongs to the blowing up of R with center I if and only if there is some $x \in I$, $x \neq 0$, some prime ideal P in $R[I/x]$ and an R-isomorphism $R[I/x]_P \to R_1$ making the diagram

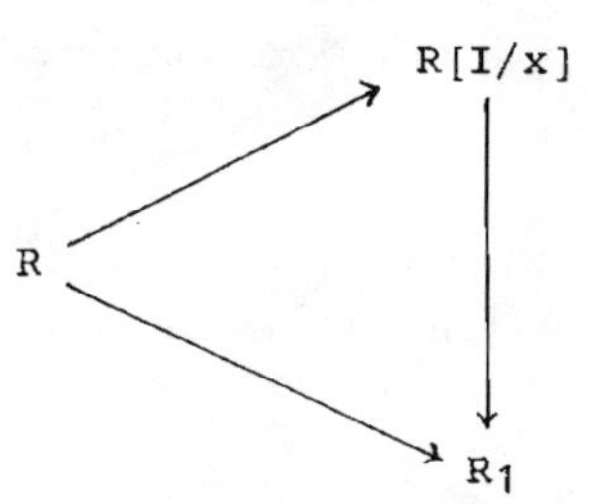

commutative, where $R \to R[I/x]_P$ is the canonical homomorphism.

Let $R_1 = R[I/x]_P$ for some nonzero x in I and some prime ideal P of $R[I/x]$. If $a \in I$, then $a/1 = x \cdot (a/x)$ in $R[I/x]$, therefore $I \cdot R[I/x] = x \cdot R[I/x]$ and of course $I \cdot R_1 = xR_1$. We note that x is a non-zerodivisor in R_1 since it is a non-zerodivisor in $R[I/x]$. If R is a domain (which will be the case in the applications later on), R_1 will be considered to be a subring of the quotient field of R, so the homomorphism $R \to R_1$ is simply the inclusion.

Let $R_1 = R[I/x]_P$ as above and let Q be the inverse image of P in R. If P does not contain the image of x in $R[I/x]$, or equivalently if Q does not contain I, then R_1 is R-isomorphic to R_Q. To see this, observe that R_x maps canonically to R_Q, and the image of $R[I/x]$ under this map is the same as the canonical image of R. So the more interesting case is the one in which $IR_1 = xR_1$ is contained in $M(R_1)$. These rings can also be described by using associated graded rings (see §3).

If R_1 is obtained by blowing up R with center I and if J is an ideal of R contained in I, one can define an ideal J_1 in R_1, called the strict transform of J in R_1, with the following property: J_1 contains JR_1, and the induced homomorphism $R/J \to R_1/J_1$ isobtained by blowing up R/J with center I/J. J_1 is uniquely determined by these properties. In the applications we will only need the case where J is a nonzero principal ideal, R is a regular local ring and I an ideal such that R/I is regular again. In this case, for any nonzero $f \in I$ let n be the largest integer such that $f \in I^n$ and put $f_1 = f/x^n$. Then f_1R_1 will be called the strict transform of fR in R_1. One checks easily that f_1R_1 does not depend on the choice of the element $x \in I$ for which $I \cdot R_1 = xR_1$. Sometimes we will also call f_1 the strict transform of f, although f determines f_1 only up to a unit. Finally one can check that f_1R_1 has indeed the universal property for the strict transform stated above. (To see this, note that $f \cdot g \in I^m$ implies $g \in I^{m-n}$ for any $g \in R$, since R and R/I were assumed to be regular.)

If I is generated by elements $x_1, \ldots, x_s$ and if R_1 is obtained by blowing up R with center I, then it is easy to see that R_1 is R-isomorphic to a localization of $R[I/x_j]$ for some j, $1 \leq j \leq s$.

§3 Associated graded rings

For any ring R and any ideal I in R, the associated graded ring of R with respect to I is defined to be

$$gr_I(R) = \bigoplus_{n \geq 0} I^n/I^{n+1}.$$

If R is local, the associated graded ring with respect to the maximal ideal will simply be denoted by gr(R). If $f \in R$, the initial form $in_I(f)$ of f with respect to I is the canonical image of f in I^n/I^{n+1}, where n is the largest integer such that $f \in I^n$ ($in_I f = 0$ if $n = \infty$). If R is local, then the initial form of f with respect to M(R) will be denoted by in(f). We will also have occasion to look at the initial ideal of an ideal J of R where R is local. This is the ideal in gr(R) generated by the initial forms of elements of J (with respect to M(R)), and it will be denoted by gr(J,R).

To describe the connection between blowing up and associated graded rings we introduce some notation. Let $S = \bigoplus_{n \geq 0} S_n$ be a graded ring and J a homogeneous ideal. The homogeneous part of degree n of J will be denoted by J_n. A homogeneous prime ideal Q of S will be called relevant if $Q_1 \neq S_1$. By $S_{(Q)}$ we denote the subring of S_Q consisting of quotients a/b, where a,b are homogeneous elements of the same degree in S and $b \notin Q$. If Q is relevant, then as for ordinary localization, the prime ideals of $S_{(Q)}$ correspond to the homogeneous prime ideals of S contained in Q in a one-to-one manner.

Now let R be a local ring, I an ideal of R, x a nonzero element of I and P a prime ideal of R[I/x]. Let $R_1 = R[I/x]_P$ and assume that $x \cdot R_1 \subset M(R_1)$. We claim that there is a uniquely determined relevant homogeneous prime ideal Q of $gr_I(R)$ such that R_1/xR_1 and $gr_I(R)_{(Q)}$ are isomorphic as R-Algebras. To see this, take Q to be the ideal generated by the initial forms $in_I(y)$, where $y \in I^n \setminus I^{n+1}$ for some n and $y/x^n \in P$, or more precisely Q_n may be defined to consist of all $in_I(y)$ where $y \in I^n$ and $y/x^n \in P$. Then Q is a prime ideal, and since we may assume $x \in I \setminus I^2$, $in_I(x) \notin Q$ and Q is relevant. Now it is clear how to define a homomorphism from $gr_I(R)_{(Q)}$ to R_1/xR_1. If $a/b \in gr_I(R)_{(Q)}$, where a and b are of degree n, let $a = in_I(y)$ and $b = in_I(z)$ for some $y,z \in I^n$. Then $b \notin Q$ implies $z/x^n \notin P$, and therefore the residue class of $y/z = (y/x^n)/(z/x^n) \bmod xR_1$ will be defined to be the image of a/b. It is left to the reader to verify that this defines indeed an isomorphism.

For the application of blowing up to resolution of surfaces in chapter III it is enough to look at those rings $R_1 = R[I/x]_P$ for which P is a maximal ideal of R[I/x]. Writing $R_1/xR_1 \simeq gr_I(R)_{(Q)}$ as above, this means that Q is a maximal element in the set of relevant homogeneous primes of $gr_I(R)$. Under suitable assumptions one can show that this implies that $Q_0 = M(R)/I$. This is easy to see for the application we have in mind, namely the case in which R is a regular local ring of dimension 3 and R/I is a onedimensional regular local ring. In this case $gr_I(R)$ is (isomorphic to) a polynomial ring in two variables over R/I, and if Q is a maximal homogeneous relevant prime in this ring, then ht(Q) = 2 and consequently $Q_o = M/I$.

Now if $Q_o = M/I$, then Q corresponds to a unique relevant homogeneous prime $\bar{Q}$ of $gr_1(R) \otimes_R R/M(R)$, and

$$R_1/M(R)R_1 \simeq (gr_I(R) \otimes_R R/N(R))_{(\bar{Q})}$$

as R-algebras. Note that under the assumptions made above, $gr_I(R) \otimes_R R/M(R)$ is a polynomial ring in two variables over R/M(R) and therefore $\bar{Q}$ is a principal ideal.

It should be clear from these considerations that the object of study for the resolution of surfaces (and also of curves) will be local homomorphisms $R \to R_1$, where $\dim R_1 = \dim R$.

<u>Definition.</u> A local homomorphism $R \to R_1$ of local rings will be called a quadratic transform, if this homomorphism is obtained by blowing up R with center M(R) and $\dim R_1 = \dim R$.

<u>Definition.</u> A local homomorphism $R \to R_1$ of local rings will be called monoidal transform with center P if this homomorphism is obtained by blowing up R with center P such that $P \neq M$ and R/P is regular, and $\dim R_1 = \dim R$.

To conclude this section we remark that any quadratic (resp. monoidal) transform $R \to R_1$ of R induces uniquely a quadratic (resp. monoidal) transform of the completion R^* of R. As noted above, such a transform is uniquely determined by a (relevant) homogeneous prime ideal Q of gr(R) (resp. $gr_P(R) \otimes_R R/M(R)$), and these graded rings do not change when passing to completion.

§4 Some special results

Now let R be a regular local ring and P a prime ideal of R such that R/P is regular, and let us look a little closer at the rings obtained by blowing up R with center P. Now R has a regular system of parameters $x_1, \ldots, x_r, x_{r+1}, \ldots, x_d$ such that P is generated by $x_1, \ldots, x_r$. Let $t = x_1$ and let $R_1 = R[P/t]_N$ for some maximal ideal N of R[P/t].

Then R_1/tR_1 is isomorphic to $gr_P(R)_{(Q)}$ for some homogeneous prime ideal Q of $gr_P(R)$ such that $in_P(x_1) \notin Q$. Let $X_i = in_P(x_i)$, $i = 1, \ldots, r$, so that $gr_P(P)$ can be identified with the polynominal ring $(R/P)[X_1, \ldots, X_r]$. Then $gr(R)_{(Q)}$ is a localization of the polynominal ring $(R/P)[X_2/X_1, \ldots, X_r/X_1]$ and hence regular. Consequently R_1 is regular and x_1 is part of a regular system of parameters of R_1.

We will describe a regular system of parameters explicitely in the case that $\dim R = \dim R_1 = 3$. Assume first that $P = (x_1, x_2)$. In this case $R_1/M(R)R_1 = R_1/(x_1,x_3)R_1$ is isomorphic to a localization of the polynomial ring $R/M(R)[X_2/X_1]$ at a principal ideal, generated by $\bar{p}(X_2/X_1)$ say. Let $p(T) \in R[T]$ be obtained by lifting the coefficients of $\bar{p}$ to R. Then $R_1/(x_1, p(x_2/x_1), x_3)R_1$ is a field and hence $x_1, p(x_2/x_1), x_3$ is a regular system of parameters. Note that $\bar{p}$ is just the minimal polynomial of the residue of X_2/X_1 mod $M(R_1)$ over $R/M(R)$ (up to a unit).

Next assume that $P = (x_1, x_2, x_3) = M(R)$, and let $R_1/tR_1 \simeq R/M(R)[X_2/X_1, X_3/X_1]_M$, where M is a maximal ideal of $R/M(R)[X_2/X_1, X_3/X_1]$. It is now clear that M is generated by two polynomials $\bar{p}_2, \bar{p}_3$, the first of which may be chosen to be the minimal polynomial of the residue of x_2/x_1 mod $M(R_1)$ over $R/M(R)$. If p_i are obtained by lifting the coefficients of $\bar{p}_i$ to R $(i = 2,3)$, then clearly $x_1, p_2(x_2/x_1), p_3(x_3/x_1)$ will be a regular system of parameters of R_1.

We will need the following result related to the strict transform of an element in R. Assume that R is a regular local ring and that R_1 is a monoidal or quadratic transform of R with center P (so in particular R/P is regular). We assume that $ht(P) = 2$. Let f be a nonzero element of R and f_1 a strict transform of f in R_1. We claim that if $ord(R)(f) = ord(R_1)(f_1) > 0$, then R_1 is uniquely determined and the homomorphism $R \to R_1$ induces an isomorphism of residue fields. To prove this, we fix the following notation. Let $n = ord(R)(f)$ and $P = (x,y)R$. Then $gr_P(R) \otimes_R R/M(R)$ can be identified with $R/M(R)[X,Y]$, and $R_1/M(R) \cdot R_1$ is isomorphic to $R/M(R)[X,Y]_{(Q)}$, Q a relevant homogeneous prime ideal. Let g be the image of $in_P(f)$ in $gr_P(R) \otimes_R R/M(R)$. Then g is homogeneous of degree $\leq n$ and $ord(R_1)(f_1) = n$ implies that $g \in Q^{(n)}$. But under our assumption Q must be principal. It follows that Q is generated by a linear form h and $g = h^n$ (up to a unit). This shows uniqueness of Q and hence of R_1, and since h is of degree 1, $R_1/M(R_1) = R/M(R)$. Note that g must be of degree n, i.e. $f \in P^n$, and therefore g may be identified with $in(f) \in gr(R)$ via the canonical homomorphism $gr_P(R) \otimes_R R/M(R) \to$

gr(R). We conclude with the following remark. Assume that $PR_1 = xR_1$. Since $R_1/M(R)R_1 = R/M(R)$, we have $(y/x)-a \in M(R_1)$ for a suitable $a \in R$. By the considerations made earlier it is clear that $x, (y/x)-a$ will be part of some regular system of parameters of R_1.

II Resolution of plane curves by weighted initial forms

Throughout this chapter, R will denote a two-dimensional regular local ring. We put $M = M(R)$, $k = R/M$, and for any $a \in R$, $p(a)$ denotes the residue class of a in k.

§1 Weighted initial form with respect to a fixed regular system of parameters

Proposition 1. Let x, z be a regular system of parameters and let v, w, e be positive real numbers. Assume that there are elements $a(i,j)$ in R $(0 \leq i,j \leq n)$ such that

$$\sum_{iv+jw=e} a(i,j)x^i z^j = \sum_{iv+jw>e} a(i,j)x^i z^j$$

Then $a(i,j) \in M$ for all (i,j) such that $iv+jw=e$.

Proof. Let

$$F(X,Z) = \sum_{iv+jw=e} a(i,j)X^i Z^j \tag{1.1}$$

and

$$G(X,Z) = \sum_{iv+jw>e} a(i,j)X^i Z^j. \tag{1.2}$$

If $a \in M$, say $a = bx+cz$, then $ax^i z^j = bx^{i+1}z^j + cx^i z^{j+1}$, so to prove the assertion we may assume in addition that

$$\left.\begin{array}{l}\text{for all } (i,j) \text{ such that } iv+jw=e \text{ we have that either}\\ a(i,j) = 0 \text{ or } a(i,j) \text{ is a unit in R.}\end{array}\right\} \tag{1.3}$$

Put $s(j) = e/v - j(w/v)$. Then (1.1) can be rewritten as

$$F(X,Z) = \sum_{t=0}^{d} c(t)X^{s(t)}Z^t, \tag{1.4}$$

where again for all t

$$c(t) = 0 \text{ or } c(t) \text{ is a unit in R.} \tag{1.5}$$

We have to show that $F(X,Z) = 0$, so assume the contrary and choose d in (1.4) such that $c(d) \neq 0$. Now let

$$F_o(X,Z) = F(X,Z) - c(d)X^{s(d)}Z^d = \sum_{t=0}^{d-1} c(t)X^{s(t)}Z^t \tag{1.6}$$

and

$$G(X,Z) = G_1(X,Z) + G_2(X,Z) \tag{1.7}$$

where

$$\left.\begin{aligned} G_1(X,Z) &= \sum_{\substack{iv+jw>e \\ j\leq d}} a(i,j)X^iZ^j \\ G_2(X,Z) &= \sum_{\substack{iv+jw>e \\ j>d}} a(i,j)X^iZ^j. \end{aligned}\right\} \tag{1.8}$$

Then we can write $G_2(X,Z) = Z^{d+1}H(X,Z)$ and we get

$$c(d)\cdot x^{s(d)}z^d - z^{d+1}H(x,z) = G_1(x,z) - F_o(x,z). \tag{1.9}$$

We have

$$\left.\begin{aligned} &s(t) > s(d) \text{ if } t < d \text{ and} \\ &j \leq d \text{ and } iv+jw > e \Rightarrow i > s(d). \end{aligned}\right\} \tag{1.10}$$

From this we see that

$$F_o(x,z) \in x^{s(d)+1}R \text{ and } G_1(x,z) \in x^{s(d)+1}R.$$

Therefore

$$z^d(c(d)x^{s(d)} - zH(x,z)) \in x^{s(d)+1}R$$

by (1.9), and we conclude that

$$c(d)x^{s(d)} - zH(x,z) \in x^{s(d)+1}R.$$

This clearly implies $c(d) \in M$ in contradiction to (1.5) and $c(d) \neq 0$.

Remark. This proof shows that the same result is valid if x,z is a regular sequence in any local ring R and M is replaced by $xR + yR$.

For the rest of this section we fix a regular system of parameters x,z of R and a nonzero element f of R, and we put $n = \mathrm{ord}(R)(f)$. We define

$$V(z,x;f) = \{v \in \mathbb{R} \mid v \geq 1 \text{ and } f = \sum_{iv+j\geq nv} a(i,j) z^i x^j \text{ for some } a(i,j) \in \mathbb{R}\}.$$

Given $v \in V(z,x;f)$ and $a(i,j) \in R$ such that

$$f = \sum_{iv+j\geq nv} a(i,j) z^i x^j$$

we define $w = \min\{j/(n-i) \mid i < n \text{ and } a(i,j) \neq 0\}$. Then $w \geq v$, $w \in (1/n!)\mathbb{Z}$ and $w \in V(z,x;f)$. It follows that

$$\sup V(z,x;f) \in (1/n!)\mathbb{Z} \cup \{\infty\}.$$

Let $y \in R$ be such that $zR+yR = M$ and let us compare $V(z,x;f)$ and $V(z,y;f)$. First of all $x = az+by$ for some unit b of R. If $v \in V(z,x;f)$ and

$$f = \sum_{iv+j\geq nv} a(i,j) z^i x^j,$$

by substituting $x = az+by$ we get an equation of the form

$$f = \sum_{r,s} b(r,s) z^r y^s,$$

and we have

$$\{(r,s) \mid b(r,s) \neq 0\} \subseteq \{(i+j-k,k) \mid iv+j\geq nv \text{ and } j\geq k\}.$$

Therefore if $b(r,s) \neq 0$, it follows that

$$rv + s = iv + jv - kv + k = iv + j + (v-1)(j-k) \geq nv$$

since $v \geq 1$ and $j \geq k$. Therefore $v \in V(z,y;f)$ and by symmetry we obtain

$$V(z,x;f) = V(z,y;f)$$

and in particular

$$\sup V(z,x;f) = \sup V(z,y;f).$$

Definition. a) $v(z;f) = \sup V(z,x;f)$.
b) $v(f) = \sup \{v(z;f) \mid z \in M \setminus M^2\}$. So $v(z;f)$ and $v(f)$ are elements of $(1/n!)\mathbb{Z} \cup \{\infty\}$.
Clearly we have $v(z;z^n) = \infty$. If r is a nonnegative integer and $ir + j \geq nr$, then

$$z^i x^j \in \begin{cases} z^n R & \text{if } i \geq n \\ M^r & \text{if } i < n \end{cases}$$

It follows that if $v(z;f) = \infty$ then $f \in \bigcap_r (z^n R + M^r) = z^n R$, and therefore

$$v(z;f) = \infty \Leftrightarrow fR = z^n R.$$

We proceed to define the weighted initial form of f with respect to z,x, which will be an element of the polynomial ring $k[Z,X]$ (remember $k = R/M$). For this purpose $k[Z,X]$ will not be identified with $gr(R)$, since the definition of the weighted initial form is not intrinsic (but see remark below).

Definition. If $v \in V(z,x;f)$ and

$$f = \sum_{iv+j \geq nv} a(i,j) z^i x^j$$

we define

$$L(z,x;v)(f) = \sum_{iv+j=nv} p(a(i,j)) Z^i X^j$$

Remark. This is well defined by Proposition 1. If z and v are chosen such that $v = v(z;f) = v(f)$, then $L(z,x;v)(f)$ can be defined intrinsically as an element of $gr(R)$ (see [8]).

Let $v = v(z;f)$ and let $w \in V(z,x;f)$ be such that $w < v$. Then we have an equation

$$f = \sum_{iv+j \geq nv} a(i,j) z^i x^j$$

and for $i \neq n$ we have $iw + j > nw$. Therefore

$$L(z,x;w)(f) = p(a(n,0))\cdot Z^n.$$

Conversely assume that $w \in V(z,x;f)$ is such that

$$L(z,x;w)(f) = p(a)Z^n, \quad a \in R.$$

Then

$$f = az^n + \sum_{iw+j>nw} a(i,j)z^i x^j,$$

and we define

$$u = \inf \{j/(n-i) \mid a(i,j) \neq 0 \text{ and } i < n\}.$$

We get that

$$u > w \text{ and } u \in V(z,x;f).$$

Therefore, for any $w \in V(z,x;f)$ we have

$$w < v(z;f) \Leftrightarrow L(z,x;w)(f) = p(a)Z^n \text{ for some } a \in R. \tag{1.1}$$

§2 Criterion for $v(z;f) = v(f)$

Proposition 2. Let z^*,x^* be a regular system of parameters of R, let f be a nonzero element of R, $n = \mathrm{ord}(R)(f)$, and let

$$F(Z,X) = L(z^*,x^*;v(z^*;f))(f) \in k[Z,X].$$

Then $v(z^*;f) < v(f)$ if and only if there is a polynomial $G(Z,X) \in k[Z,X]$ and an element $a \in R$ such that

$$F(Z,X) = p(a)\ G(Z,X)^n.$$

The proof of this proposition is the content of this section and will be obtained by an explicit description of the weighted initial form for a second system of parameters z,x of R.

Let $v \in V(z,x;f)$. Then f can be written as

$$f = az^n + \sum_{\substack{iv+j\geq nv \\ i<n}} a(i,j)z^i x^j. \qquad (2.1)$$

Let w be a positive integer and $z^* = z - bx^w$, so that z^*,x generate M. Then

$$f = a(z^*+bx^w)^n + \sum_{\substack{iv+j\geq nv \\ i<n}} a(i,j)\sum_{k=0}^{i}\binom{i}{k}b^k y^{i-k}x^{j+kw} \qquad (2.2)$$

Note that $(i-k)w + (j+kw) = iw+j$. If now $w < v$ and $i < n$, then

$$iv + j \geq nv \Leftrightarrow j \geq (n-i)v \Rightarrow j > (n-i)w \Leftrightarrow iw + j > nw.$$

Therefore, if $w < v$ then

$$L(z^*,x;w)(f) = L(z^*,x;w)(a(z^*+bx^w)^n) = p(a)(Z + p(b)X^w)^n. \qquad (2.3)$$

From (1.1) we conclude that $v(z^*;f) = w < v$ if $p(a) \neq 0 \neq p(b)$. Therefore

$$\left.\begin{array}{l}\text{If } f \text{ is as in (2.1) with } a \notin M \text{ and } z^* = z - bx^w \text{ with} \\ b \notin M, \text{ then } v(z^*;f) \geq v \Rightarrow w \geq v.\end{array}\right\} \qquad (2.4)$$

Let us rewrite (2.2) as

$$f = \sum_{iv+j\geq nv} a(i,j)\sum_{k=0}^{i}\binom{i}{k}b^k y^{i-k}x^{j+kw} \qquad (2.5)$$

where $a(n,0) = a$. Now

$$(i-k)v + j + kw = dv \Leftrightarrow \begin{cases} iv+j = dv \text{ and } k = 0 & \text{if } w > v \\ iv+j = dv & \text{if } w = v.\end{cases}$$

Therefore

$$L(z^*,x;v)(f) = \begin{cases} \displaystyle\sum_{iv+j=nv} p(a(i,j))Z^iX^j & \text{if } w>v \\ \displaystyle\sum_{iv+j=nv} p(a(i,j))\sum_{k=0}^{i}\binom{i}{k} p(b)^k Z^{k-i}X^{j+kv} & \text{if } w=v.\end{cases} \qquad (2.6)$$

Define

$$G(Z,X) = L(z^*,x;v)(z) = \begin{cases} Z & \text{if } w > v \\ Z + p(b)X^v & \text{if } w = v \end{cases}$$

and define $H(Z,X) = L(z,x;v)(f)$. Then (2.6) can be written as

$$L(z^*,x;v)(f) = H(G(Z,X),X). \tag{2.7}$$

Furthermore in the case $v = v(z;f)$, by (1.1) and (2.7) we have

$$w > v(z;f) \Rightarrow v(z^*;f) = v(z;f). \tag{2.8}$$

We now turn to the proof of Proposition 2 and assume that we are given regular systems of parameters z,x and z^*,x^* of R such that

$$v(z^*;f) < v(z;f). \tag{2.9}$$

Assume first that $v(z^*;f) = 1$. Writing

$$f = \sum_{i+j=n} a(i,j)\, z^{*i} x^{*j}$$

we see that in $gr(R)$

$$\text{in } f = \sum_{i+j=n} p(a(i,j))\, \text{in}(z^*)^i \text{in}(x^*)^j = p(a)\cdot \text{in}(z)^n$$

for some $a \in R$. So assume $v(z^*;f) > 1$. Then

$$\text{in } f = p(a)\cdot \text{in}(z)^n = p(a')\,\text{in}(z^*)^n, \quad a,a' \in R$$

and therefore z^*,x generate M. Let $w = v(z^*;z)$. Then

$$z = cz^* + bx^w$$

and c,b are units in R. We may assume $c = 1$. Let $v = v(z^*;f) \in V(z,x;f)$ and $H(Z,X) = L(z,x;v)(f)$. By (1.1) and (2.9), $H(Z,X) = p(a)\cdot Z^n$ for some $a \in R$, and so from (2.3) and (2.7) we see that

$$L(z^*,x;v)(f) = p(a)\cdot G(Z,X)^n, \quad v = v(z^*;f) \tag{2.10}$$

where $G(Z,X) = L(z^*,x;v)(z)$.

Assume conversely that (2.10) holds. Then necessarily we have

$$G(Z,X) = p(c)Z + p(d)X^v.$$

So if we define $z = cz^* + dx^v$, from (2.7) we conclude that

$$H(Z,X) = L(z,x;v)(f) = p(a)Z^d$$

and therefore $v(z;f) > v$ by (1.1).

§3 Criterion for $v(f) = \infty$

Let $f \in R$, $\mathrm{ord}(R)(f) = n < \infty$ as before and let R^* be the completion of R. If z,x is a regular system of parameters of R and $v \in V(z,x;f)$, then for the weighted initial form $L(z,x;v)(f)$ it does not make any difference whether f is considered as element of R or as element of R^*. Consequently, by (1.1) and Proposition 2, the numbers $v(z;f)$ and $v(f)$ will be the same for R and R^*.

Assume that $v(f) = \infty$, and let $z_0 = z$, $v_0 = v(z_0;f)$. If $v_0 < \infty = v(f)$ then by the proof of Proposition 2 we can find e_1, $c_1 \in R$ so that

$$v_1 = v(z_1;f) > v_0 \text{ where } z_1 = e_1 z_0 + c_1 x^{v_0}.$$

Proceeding inductively, we either obtain some $z_{i-1} \in R$ such that $fR = z^n_{i-1}R$, or we can find

$$v_i \in \mathbb{N},\ e_i \in R,\ c_i \in R \text{ and } z_i \in R$$

such that

$$z_i = e_i z_{i-1} + c_i x^{v_{i-1}},\ v_i = v(z_i;f) > v_{i-1} = v(z_{i-1};f)$$

Actually we may assume $e_i = 1$ for all i, so that

$$z_{i+1} - z_i \in M^{v_i} \text{ and } \lim_{i\to\infty} v_i = \infty.$$

Let

$$z^* = \lim_{i\to\infty} z_i \in R^*.$$

Then

$$z^* - z^i \in M^{v_i} \quad \text{for all } i$$

and therefore

$$f \in (z^*R^*)^n + (MR^*)^{v_i} \quad \text{for all } i.$$

Therefore

$$f\,R^* = (z^*R^*)^n.$$

So we have shown

Proposition 3. $v(f) = \infty$ if and only if there is a regular parameter z^* of R^* such that $fR^* = (z^*R^*)^n$.

Assume in addition that R is excellent and let $P = \sqrt{fR}$. Then $P\cdot R^* = \sqrt{fR^*}$. Therefore, if $v(f) = \infty$, then $R^*/PR^* = (R/P)^*$ is regular and P is generated by some regular parameter z of R. So we get

Proposition 3'. If R is excellent, then $v(f) = \infty$ if and only if there is a regular parameter z of R such that $fR = z^nR$.

§4 Immediate quadratic transforms

Let f be a nonzero element of R such that $v(f) < \infty$. We choose a regular system of parameters z,x of R such that $v(f) = v(z;f)$. We let $n = \operatorname{ord}(R)(f) > 0$ as before. Let $R \to R_1$ be a quadratic transform of R and let f_1 be a strict transform of f in R_1, and assume $\operatorname{ord}(R_1)(f_1) = n$. By the observation made in I,§4, R_1 is uniquely determined and in(f) is equal (up to a unit) to the nth power of a homogeneous element of degree one in gr(R). Therefore $v(f) > 1$, and by the choice of z we have $\operatorname{in}(f) = \operatorname{in}(z)^n$ up to a unit. Hence $z_1 = z/x$, x is a regular system of parameters of R_1, and $R_1/M(R_1) = k$. Let $v_1 = v(z_1;f_1)$ and write

$$f_1 = \sum_{\substack{iv_1+j\geq nv_1\\ i\leq n}} b(i,j)\, z_1^i x^j \quad \text{with } b(i,j) \in R_1$$

Then

$$f = f_1 x^n = \sum_{\substack{iv_1+j\geq nv_1\\ i\leq n}} b(i,j)\, z^i x^{j+n-1},$$

and

$$i(v_1+1) + j + n - i = iv_1 + j + n \geq n(v_1+1).$$

Since $R_1/M(R_1) = R/M$, we may assume that

$$b(i,j) \in R \text{ if } iv_1 + j = nv_1.$$

It follows that

$$v(f) \geq v_1 + 1 \geq 2$$

and furthermore if we put

$$L(z,x;v_1+1)(f) = F(Z,X)$$
$$L(z_1,x;v_1)(f) = F_1(X,Z)$$

then

$$F_1(Z,X) = X^n F(Z/X,X).$$

From (1.1) we conclude that $v_1 + 1 = v(f)$, and therefore by Proposition 2 it follows that $v_1 = v(f_1)$.

Now instead of $\text{ord}(R_1)(f_1) = n$, assume

$$\text{ord}(R_1)(f_1) > 0 \text{ and } v(f) \geq 2.$$

Then $\text{in}(f) = p(a)\cdot \text{in}(z)^n$ for some $a \in R$, so R_1 is the only quadratic transform of R for which $\text{ord}(R_1)(f_1) > 0$, and $z_1 = z/x$, x is a regular system of parameters of R_1. Writing

$$f = \sum_{\substack{2i+j\geq 2n \\ i\leq n}} a(i,j) z^i x^j$$

where $v = v(f)$, we get

$$f_1 = \sum_{\substack{2i+j\geq 2n \\ i\leq n}} a(i,j) z_1^i x^{j+i-n}$$

we see that $f_1 \in M(R_1)^n$. We summarize our results in

Proposition 4. Let $f \in R$ be such that $\mathrm{ord}(R)(f) = n < \infty$ and $v(f) < \infty$. Choose $z,x \in R$ such that $M = zR + xR$ and $v(f) = v(z;f)$. Let $R \to R_1$ be a quadratic transform of R and f_1 a strict transform of f in R_1, and assume $f_1 \in M(R_1)$. Then $\mathrm{ord}(R_1)(f_1) = n$ if and only if $v(f) \geq 2$, and in this case we have $z_1 = z/x \in M(R_1)$ and

$$v(f_1) = v(z_1;f_1) = v(f) - 1.$$

§5 Embedded resolution of plane curves and principalization

For this section we fix a sequence

$$R = R_o \to R_1 \to R_2 \to \dots \to R_i \to \dots$$

of two-dimensional regular local rings and elements $f_i \in R_i$ such that

$$\left.\begin{array}{l} R_i \text{ is a quadratic transform of } R_{i-1} \\ f_i \text{ is a strict transform of } f_{i-1} \end{array}\right\} \text{ for all } i \geq 1.$$

We put $f = f_o$ and assume, of course, that $0 < \mathrm{ord}(R)(f) < \infty$. By Proposition 3' and Proposition 4 we know:

$$\left.\begin{array}{l} \text{If } R \text{ is excellent, then there is some } i_o \text{ and for} \\ \text{each } i \geq i_o \text{ there are } z_i \in M(R_i) \setminus M(R_i)^2 \text{ and a} \\ \text{nonnegative integer } t_i \text{ such that } f_iR_i = (z_iR_i)^{t_i}. \end{array}\right\} \qquad (5.1)$$

If z,x is a regular system of parameters of R such that $M(R)R_1 = xR_1$, then either z/x is a unit in R_1 or $x, z/x$ is a regular system of parameters in R_1. Using this, one shows by induction that

$$(f/f_i)\cdot R_i \text{ has normal crossings for all } i \geq 0. \qquad (5.2)$$

Using the notation of (5.1), choose $i \geq i_o$ and extend z_i to a regular system of parameters $z_i = z,x$ of R_i. Furthermore let u,v be a regular system of parameters of R_i such that

$$(f/f_i)R_i = u^a v^b R_i \qquad (5.3)$$

for some integers a,b. Assume that $M(R_i)R_{i+1} = uR_{i+1}$. By (5.1) and (5.3) we have

$$f R_i = z^t u^a v^b R_i \quad \text{where } t = t_i.$$

If $M(R_i)R_{i+1} = zR_{i+1}$, then $fR_{i+1} = u^{a+b+t}(v/u)^b R_{i+1}$ has normal crossings. If $M(R_i)R_{i+1} = xR_{i+1}$, then

$$f R_{i+1} = (z/x)^t \, x^{a+b+t} \, (v/u)^b \, R_{i+1}, \tag{5.4}$$

and this has normal crossings again, if z/x or v/u is a unit in R_{i+1}. So let us look at the case that $z/x \in M(R_{i+1})$ and $v/u \in M(R_{i+1})$. Then z/x, x is a regular system of parameters of R_{i+1}. We define

$$w_o(a) = \mathrm{ord}(R_i/zR_i)(a + zR_i) \quad \text{for } a \in R_i,$$
$$w_1(b) = \mathrm{ord}(R_{i+1}/(z/x)R_{i+1})(b + (z/x)R_{i+1}) \quad \text{for } b \in R_{i+1}.$$

Then

$$w_o(u) > 0 \text{ and } w_o(v) = w_1(v).$$

Therefore

$$w_1(u\cdot(v/u)) = w_1(v) < w_o(u\cdot v).$$

It follows that after finitely many steps we obtain an equation like (5.4) in which either z/x or v/u is a unit. This proves

Proposition 5. Assume that R is excellent. Given $f \neq 0$ in R and any sequence

$$R = R_o \to R_1 \to \dots \to R_i \to \dots$$

of successive quadratic transforms, there is an integer i_1 such that fR_i has normal crossings for all $i \geq i_1$.

Given nonzero elements $g_1,\dots,g_m$ in R and applying Proposition 5 to $f = g_1 \dots g_m$ we obtain the following

Corollary. Assume that R is excellent and

$$R = R_o \to R_1 \to \dots \to R_i \to \dots$$

is a sequence of successive quadratic transforms. Given any nonzero elements $g_1,\dots,g_m$ of R, there is an integer i_1 such that the set

$$\{g_1R_i, \ldots, g_mR_i\}$$

has normal crossings for all $i \geq i_1$.

Proposition 6. Assume that R is excellent and that

$$R = R_o \to R_1 \to \ldots \to R_i \to \ldots$$

is a sequence of successive quadratic transforms. For any ideal I of R there is an integer i_2 such that IR_i is principal for $i \geq i_2$.

For the proof we make induction on the number of generators of I, so we may assume $I = fR + gR$, $f \neq 0 \neq g$. By the corollary to Proposition 5 we can choose i and a regular system of parameters z,x of R_i such that

$$fR_i = x^a z^b R_i, \; g = x^c z^d R_i.$$

We put $t_i = (a-c)(b-d)$ and we observe that

IR_i is principal if and only if $t_i \geq 0$.

Assume $t_i < 0$ and $M(R_i)\cdot R_{i+1} = xR_{i+1}$. If z/x is a unit in R_{i+1}, then

$$IR_{i+1} = x^{a+b}R_{i+1} + x^{c+d}R_{i+1},$$

which is principal. If $z/x \in M(R_{i+1})$, we have

$$fR_{i+1} = x^{a+b}(z/x)^bR_{i+1}, \; gR_{i+1} = x^{c+d}(z/x)^dR_{i+1}$$

and therefore

$$t_{i+1} = (a+b-c-d)(b-d) = (b-d)^2 + t_i > t_i,$$

and this finishes the proof.

III Resolution and Embedded Resolution of Embedded Surfaces in Characteristic 0

§1 General assumptions and statements of the results

By a surface we will mean a noetherian scheme F, of pure dimension 2, such that

the local ring at any point of F is a factor of an excellent regular local ring containing a field of characteristic 0. (1.1)

Also in what follows, any regular local ring will be assumed to be excellent and to contain a field of characteristic 0. A surface F will be called locally embedded, if for any closed point $P \in F$, the local ring of F at P is the homomorphic image of a 3-dimensional regular local ring. F will be called embedded, if it is (isomorphic to) a closed subscheme of a regular (excellent) scheme of pure dimension 3. The purpose of this chapter is to give the main computations for the proof of the following two theorems:

Theorem A. If F is a reduced, locally embedded surface, there is a surface F^* and a morphism $\varphi: F^* \to F$ such that

a) F^* is regular.

b) φ is the composition of quadratic and monoidal transformations.

Theorem B. If F is a reduced, embedded surface, closed subscheme of the 3-dimensional, regular excellent scheme Z, there is a 3-dimensional, regular (excellent) scheme Z^* and a morphism $\varphi: Z^* \to Z$ such that

a) φ is the composition of quadratic and monoidal transformations.

b) The (iterated) strict transform F^* of F is regular.

c) $\varphi^{-1}(F)$ has normal crossings.

(For the general algebraic definition of normal crossings see I,§1). In fact it turns out that the quadratic and monoidal transformations used for φ are of a certain restricted type; e.g. for Theorem A, the center of such a transformation is contained in the singular locus of the surface in question. Therefore F^* contains an open set U such that φ maps U isomorphically to the set of regular points of F.

The proofs of both theorems consist mainly of local computations, the global argument being reduced to a certain finiteness statement (see

§5 and §9). In this chapter we give the local computation in the complete case. The results needed to reduce the general case to the complete one, together with other technical details, are the content of the next chapter. Reviews of the method of proof for Theorems A and B will be given later on (see §6 and §10), but we will anticipate two points of these reviews in order to give a small background for the following algebraic setup.

To prove theorem A, it is obviously sufficient to obtain a morphism $\varphi: F^* \to F$ with b) and such that F^* has the following property: If $e(F)$ denotes the highest multiplicity which F has in its (closed) points, then $e(F^*) < e(F)$. For the proof of theorem B let us look at any morphism $\varphi: Z^* \to Z$ which is the composition of quadratic and (suitable) monoidal transformations. Then $\varphi^{-1}(F) = F^* \cup \tilde{F}$, where F^* is the (iterated) strict transform and $\tilde{F}$ the exceptional divisor. One trick of the proof is to decompose $\tilde{F}$ as $\tilde{F} = G \cup H$, where G is the "old" exceptional divisor obtained in those steps in which the maximal multiplicity of the surface (= iterated strict transform of F) was bigger than $e(F^*)$, and H is the "new" exceptional divisor of those last steps in which the maximal multiplicity did not drop. Since our surface F in the theorem is supposed to be (locally) embedded, locally we have to consider a 3-dimensional regular local ring R (subject to the assumptions made earlier) and an element $f \in R$ which defines F locally. We denote by n the number $\mathrm{ord}(R)(f)$, i.e. the multiplicity of F at the point represented by R, and we assume $n \geq 1$ and $\sqrt{fR} = fR$.

The notation R, f, n will be kept fixed for this chapter, and all the assumptions made before will be in force.

§2 Main example: Strict transform for $f = z^n + x^a y^b$

The reason to look at this example is that the general proof consists of a reduction to this example. Therefore this special case shows in the most purified way the idea, and the simplicity, of the general proof. In the following description we use some facts, which will be proved later on, although they are standard for those people who are familiar with the subject.

We will view $f = z^n + x^a y^b$ as an element of the power series ring $R = K[[x,y,z]]$, K a field of characteristic 0. The assumption n =

ord(R)(f) implies $a+b \geq n$. Let us assume $n > 1$. Then the n-fold curves are among $\{(z,x),(z,y)\}$, and (z,x) (resp. (z,y)) is n-fold if and only if $a \geq n$ (resp. $b \geq n$), see §4 for proof.

Let $a \geq n$ and $\varphi: R \to R_1$ be a monoidal transform with center (z,x), such that a strict transform f_1 of f satisfies $\mathrm{ord}(R_1)(f_1) = n$ and $\dim R_1 = 3$. Then $R_1^* = K[[x,y,z_1]]$, $z_1 = z/x$ and

$$f_1 = z_1^n + x^{a-n}y^b.$$

Now let $\varphi: R \to R_1$ be a quadratic transformation such that $\dim R_1 = 3$ and $\mathrm{ord}(R_1)(f_1) = n$, where again f_1 is a strict transform of f. Then

$$(x,y,z)R_1 = \begin{cases} xR_1 & \text{or} \\ yR_1. \end{cases}$$

If $(x,y,z)R_1 = xR_1$, then

$$R_1^* = S_1[[z_1]]$$

where

$z_1 = z/x$

S_1 = the completion of a quadratic transform of $K[[x,y]]$

and

$$f_1 = z_1^n + x^{a+b-n}(y/x)^b.$$

(y/x may be a unit in S_1). Let us write

$$a = c\cdot n + \bar{a}\ ,\ 0 \leq \bar{a} < n$$
$$b = d\cdot n + \bar{b}\ ,\ 0 \leq \bar{b} < n$$
$$r_n(a,b) = \bar{a} + \bar{b}$$

After performing $t = c + d$ monoidal transformations with an n-fold curve as center we will obtain a strict transform

$$f_t = z_t^n + x^{\bar{a}}y^{\bar{b}}.$$

If now $r_n(a,b) = \bar{a} + \bar{b} < n$, f_t will define a point on a surface of multiplicity $< n$. This is the typical example of a good point. If $r_n(a,b) \geq n$, let us apply a quadratic transformation to f like above. Then (up to units)

$$f_1 = z_1^n + x^{a_1} y_1^{b_1}, \quad y_1 = y/x$$

where

$$a_1 = a+b-n, \; b_1 = b, \text{ or } b_1 = 0.$$

If $b_1 = 0$, $r_n(a_1,b_1) < n$, and if $b_1 = b$ we get

$$r_n(a_1,b_1) = \bar{a} + \bar{b} - n + \bar{b} < r_n(\bar{a},\bar{b}).$$

So by repeated quadratic transformations we will reach the stage that $r_n(a,b) < n$.

Let us call a monoidal transformation <u>permissible</u> for f, if the center is an n-fold curve on the surface defined by f. Then we have obtained the following result: Along any sequence of quadratic transformations, there is an iterated strict transform of f for which reduction of multiplicity below n can be achieved by permissible monoidal transformations. The number of such transformations may be zero. We note that the condition $r_n(a,b) < n$ is stable under permissible monoidal transformations, but it is not stable under quadratic transformations. As an example, take $f = z^3 + x^4y^4$, which after quadratic transformation gives $z_1^3 + x^5y_1^4$ ($z_1 = z/x$, $y_1 = y/x$). This means (in the terminology to be defined in §3) that the property of a point to be good is stable under permissible monoidal transformations, but unstable under quadratic transformations.

§3 Good points: Definition and elementary properties

We start with R, f, n as before. Let us recall that for a prime ideal P (in any ring), $P^{(s)}$ denotes the sth-symbolic power of P. If P is generated by a regular sequence, then it is well known that $P^{(s)} = P^s$.

<u>Definition.</u> $E(f,R) = \{P \subset R \mid P$ prime ideal of height 2 such that $f \in P^{(n)}\}$.

So locally $E(f,R)$ is the set of n-fold curves on the surface defined by $f = 0$.

<u>Definition.</u> f is <u>pre-good</u> in R if
a) $E(f,R) \neq \emptyset$ and
b) $E(f,R)$ has normal crossings.

Let f be pre-good in R and let $\varphi: R \to R_1$ be a monoidal transformation with center $P_o \in E(f,R)$ (and $\dim R_1 = 3$), such that for a strict transform f_1 of f we have $\mathrm{ord}(R_1)(f_1) = n$. Then [1], (3.1o) contains a detailed description of $E(f_1,R_1)$ (the proof will be reproduced in the next chapter). If $Q \in E(f_1,R_1)$, then either $Q = R_1 \cap PR_P$ for some $P \in E(f,R)$, $P \neq P_o$, or

$$Q = M(S_1) \cap R_1,$$

where

S_1 = a quadratic transform of R_{P_o}
$M(S_1)$ = the maximal ideal of S_1

and

$$\mathrm{ord}(S_1)(f_1) = n.$$

Therefore $E(f_1,R_1)$ has normal crossings again, but it may be empty.

<u>Definition.</u> A monoidal transformation with center $P \subset R$ is called <u>permissible</u> for f if R/P is regular and $P \in E(f,R)$.

Now we define f to be good in R, if the property of being pre-good is stable under permissible monoidal transformations. To be precise:

<u>Definition.</u> f is <u>good</u> in R if the following conditions are satisfied:
a) f is pre-good in R.
b) Let
$$R = R_o \overset{\varphi_1}{\to} R_1 \to \dots \overset{\varphi_t}{\to} R_t$$
be any sequence of permissible monoidal transformations, let $f_o = f$ and f_i = strict transform of f_{i-1} $(i=1,\dots,t)$, and assume that $\mathrm{ord}(R_i)(f_i) = n$ for $i=1,\dots,t-1$. Then
$\mathrm{ord}(R_t)(f_t) < n$ or f_t is pre-good in R_t.

(φ_i is supposed to be permissible for f_{i-1}, of course.)

So a pre-good point which is not good has the property (by definition) that by permissible monoidal transformations it is transformed into an n-fold point which does not lie on any n-fold curve. (This will be called an isolated n-fold point.)

By localization, the permissible monoidal transformations correspond to quadratic transforms in an n-fold point of a plane curve. Therefore good points are curve-like, in the sense that for reduction of multiplicity they behave like singularities of plane curves.

§4 Strict transform: Power series case

We consider again the case $R = K[[x,y,z]]$. Then, by a suitable choice of x, y, z we may assume that

$$f = z^n + \sum_{i=2}^{n} \alpha_i z^{n-i}, \quad \alpha_i \in K[[x,y]] = S$$

and

$$\mathrm{ord}(S)(\alpha_i) \geq i, \quad 2 \leq i \leq n.$$

In S there is an element β, unique up to a unit, with the following properties:

$$\beta^i | \alpha_i \quad \text{for } i=2,\ldots,n. \tag{4.1}$$

$$\text{If } \gamma \in S \text{ and } \gamma^i | \alpha_i \text{ for } i=2,\ldots,n, \text{ then } \gamma | \beta. \tag{4.2}$$

We will use the notation

$$\beta = g \cdot c \cdot d \cdot \{\alpha_i^{1/i} \mid 2 \leq i \leq n\}. \tag{4.3}$$

Now let $P \in E(f,R)$, and let D be the partial derivative with respect to z. Then $DP^{(s)} \subset P^{(s-1)}$ for all $s \geq 1$ (as for any derivative of R), and therefore

$$D^{(n-1)}f = n!z \in P^{(1)} = P.$$

It follows that $z \in P$, and therefore $P = zR + tR$ for some prime element t of S. This implies $P^{(n)} = P^n$ and therefore

$$t^i | \alpha_i, \; 2 \leq i \leq n,$$

or

$$t | \beta$$

by (4.2). Therefore

$$E(f,R) = \{(z,t) \mid t \in S \text{ prime and } t|\beta\}. \tag{4.4}$$

From this we see that

$$f \text{ pre-good in } R \Leftrightarrow \beta \text{ is a non-unit having normal crossings,} \tag{4.5}$$

and we note that for f pre-good in R, $E(f,R)$ can have at most two elements.

Before giving a criterion for f to be good in R, let us prepare f a little. Let $\varphi: R \to R_1$ be a quadratic transformation and f_1 a strict transform such that $\mathrm{ord}(R_1)(f_1) = n$ (and $\dim R_1 = 3$). Then

$$R_1^* = S_1[[z_1]]$$

where

$$S_1 = \text{completion of a quadratic transform of } S$$
$$z_1 = z/x \quad \text{or} \quad z_1 = z/y.$$

(See next chapter for the details). This means that

$$f_1 = z_1^n + \sum_{i=2}^{n} \tilde{\alpha}_i z_1^{n-i}$$

where

$$\tilde{\alpha}_i = \tilde{\alpha}_i/x^i \quad \text{or} \quad \tilde{\alpha}_i = \tilde{\alpha}_i/y^i, \quad 2 \leq i \leq n. \tag{4.6}$$

Therefore after t quadratic transformations we get a ring R_t and an iterated strict transform f_t of f for which the following holds: If $\mathrm{ord}(R_t)(f_t) = n$, then

$R_t^* = S_t[[z_t]]$
S_t = power series ring in 2 variables

and

$$f_t = z_t^n + \sum_{i=2}^{n} \alpha_i^* z_t^{n-i}.$$

Applying resolution of plane curves to the (completed) sequence of quadratic transformations

$$S \to S_1 \to \ldots \to S_t$$

we may assume that

$\{\alpha_2,\ldots,\alpha_n\}$ has normal crossings in S_t (4.7)

and

the ideal in S_t generated by $\{\alpha_i^{n!/i} \mid 2 \leq i \leq n\}$ is principal. (4.8)

Now by (4.6) there is some $\gamma \in S_t$ such that

$$\alpha_i^* = \gamma^i \alpha_i, \ 2 \leq i \leq n,$$

and therefore the corresponding properties to (4.7) and (4.8) also hold for $\alpha_2^*,\ldots,\alpha_n^*$.

Now let us assume from the beginning that

$\{\alpha_2,\ldots,\alpha_n\}$ has normal crossings in S (4.9)

and

the ideal in S generated by $\{\alpha_i^{n!/n} \mid 2 \leq i \leq n\}$ is principal. (4.10)

Then (up to units and for a suitable choice of x,y)

$$\alpha_i = x^{a_i} y^{b_i}, \ 2 \leq i \leq n \tag{4.11}$$

and by (4.10) there is some $u \in \{2,\ldots,n\}$ such that

$$\alpha_u = g\cdot c\cdot d\cdot\{\alpha_i^{1/i} \mid 2 \leq i \leq n\} \tag{4.12}$$

and by the previous considerations, (4.11) and (4.12) are stable under quadratic transformations as long as the multiplicity of the strict transform of f is n.
But these conditions are also stable under permissible monoidal transformations. To see this, let

$$\varphi: R \to R_1$$

be a permissible monoidal transformation with center P for f such that $ord(R_1)(f_1) = n$, where f_1 is a strict transform of f. Then

$$R_1^* = S[[z_1]]$$

$$z_1 = \begin{cases} z/x & \text{if } P = (z,x) \\ z/y & \text{if } P = (z,y) \end{cases}$$

and

$$f_1 = z_1^n + \sum_{i=2}^{n} \tilde{\alpha}_i \, z_i^{n-i},$$

where

$$\tilde{\alpha}_i = \begin{cases} \alpha_i/x^i & \text{if } P = (x,z) \\ \alpha_a/y^i & \text{if } P = (z,y) \end{cases} \qquad 2 \leq i \leq n.$$

(For the description of R_1, see the next chapter again.) So we see that under the assumption (4.11) and (4.12) the behaviour of f under quadratic and permissible monoidal transformations is exactly the same as that of

$$\tilde{f} = z^u + \alpha_u = z^u + x^{a_u} y^{b_u} ,$$

and also f is good in R if and only if $\tilde{f}$ is good in R. We summarize the result as follows: Let

$$R = R_o \to R_1 \to R_2 \to \dots \to R_z \to \dots$$

be an infinite sequence of 3-dimensional regular local rings such that R_i is the completion of a quadratic transformation of R_{i-1} and let f_i be a strict transform of f_{i-1} $(i \geq 1)$. Assume that $ord(R_i)(f_i) = n$ for all $i \geq 1$. Then f_t is good in R_t for some t. For the proof we may assume f to satisfy (4.11) and (4.12), which reduces the assertion to the case $f = \tilde{f}$. But this case was treated in §2.

We emphasize that the explicit description of the behaviour of f under quadratic and permissible monoidal transformations given above is only valid under the assumption that the multiplicity remains n. No such description is possible in a step in which the multiplicity drops.

§5 Bad points are finite in number

Let us fix some notation. F will denote a surface (subject to the assumptions made in §1). By a point of F we mean a closed point, and in fact we will identify F with its set of closed points. Let $m(F) = \max\{\mathrm{mult}_P(F) \mid P \in F\}$ and $M(F) = \{P \in F \mid \mathrm{mult}_P(F) = m(F)\}$. Then M(F) is a closed subset of F, and if F is not regular, then $\dim M(F) \leq 1$. We assume from now on that F is not regular (i.e. $m(F) > 1$).

Definition. A point $P \in F$ is called good resp. pre-good if either $\mathrm{mult}_P(F) < m(F)$ or $P \in M(F)$ and the local ring of F at P is of the form R/fR with a regular local ring R of dimension 3 and f good resp. pre-good in R. (This condition does not depend on the choice of R and f.)

Granting that normal crossing of M(F) at some point P is stable under permissible monoidal transformations, we have the following geometric description of bad (= non-good) points. A point $P \in M(F)$ is bad if it is either an isolated point of M(F), or a singularity of a one-dimensional component, or a point of intersection of one-dimensional components of M(F) which do not have normal crossings at P, or finally a point that is transformed into one of these three types by permissible monoidal transformations.

To prove the finiteness of bad points, we define a subset B(F) of M(F) such that:

a) B(F) is finite.
b) All points of F-B(F) are pre-good.
c) We obtain a surface of F^* such that either $m(F^*) < m(F)$ or $M(F^*) = B(F^*)$ is finite by repeated application of the following procedure: First replace F by $F' = F-B(F)$, and if $F' \cap M(F)$ has a one-dimensional component C, let $F_1 \to F'$ be a monoidal transformation of F' with center C. Continue with F_1. (5.1)

We define

$$B(F) = \{P \in M(F) \mid P \text{ is an isolated point of } M(F) \text{ or a singular point of } M(F)\}.$$

(We note that the points of intersection of different one-dimensional components of M(F) are singular points of M(F).) Now a) and b) are clear. As noted earlier in §3, the effect of a permissible monoidal transformation on a regular point of M(F) is the same as that of a quadratic transformation on a plane curve. The means that there is a well defined number v such that after exactly v permissible transformations the corresponding point will either have a multiplicity less than m(F) or it will be an isolated m(F)-fold point. Applying this remark to all one-dimensional components of $M(F) \cap F'$ we obtain c) above. Now let

$$F^* = F_t \to F_{t-1} \to \dots \to F_1 \to F_o = F$$

be a sequence of surfaces obtained by the procedure of (5.1), c), and let $p_i: F_i \to F$ be the corresponding morphism. Then we have

If $P \in F^* - B(F^*)$, then $\text{mult}_P(F^*) < m(F)$. (5.2)

For any $P \in F^* - B(F^*)$, $p_t(P)$ is a good point of F. (5.3)

From this it is clear that the bad points of F are contained in the finite set

$$\bigcup_{i=0}^{t} p_i(B(F_i)),$$

where $p_o = id_F$.

§6 Review of the procedure to resolve F

Starting with a surface F (as defined in §1), we apply repeated quadratic transformations to F centered at the finitely many bad points of F. By this process we obtain a surface F* such that m(F*) < m(F) or every point of F* is a good point. If m(F*) = m(F), we apply monoidal transformations to F* centered at (1-dimensional) components of M(F*). We note that these centers are globally defined. Now the mere definition of a good point says that by repeating these monoidal transformations, we will eventually obtain a surface F^{**} such that $m(F^{**}) < m(F)$, and this proves Theorem A.

§7 Embedded resolution - good triples

As indicated at the end of §1, the local object of study for the proof of Theorem B consists of triples $T = (f,g,h)$ of elements of a regular local ring R, for which we make the following assumptions:

$$\left.\begin{array}{l} \sqrt{fR} = fR,\ f \neq 0 \\ gh \text{ has normal crossings} \\ f \text{ and } gh \text{ have no common prime factor} \\ n = \operatorname{ord}(R)(f) \geq 1 \end{array}\right\} \qquad (7.1)$$

We will fix this triple $T = (f,g,h)$ with (7.1) for the rest of this chapter.

T will be called resolved if $n = 1$ and $f\cdot g\cdot h$ has normal crossings. If $n = 1$, T is resolved if and only if either $g\cdot h$ is a unit or the following condition is satisfied: If $P \subset R$ is a prime ideal of height 2 such that $f \in P$ and $g\cdot h \in P$, then $f\cdot g\cdot h$ has normal crossings in R_P and R/P is regular.

<u>Definition.</u> If $n > 1$, we put $E(f,g,h;R) = E(f,R)$, and if $n = 1$ we put

$$E(f,g,h;R) = \{P \subset R \mid P \text{ prime},\ ht(P) = 2,\ f \in P,\ gh \in P \text{ and either } R/P \text{ not regular or } fgh \text{ does not have normal crossings in } R_P\}.$$

This definition is made up in such a way that T is resolved if and only if $n = 1$ and $E(f,g,h;R) = \emptyset$. The geometric meaning of $E(f,g,h;R)$ in the case $n = 1$ is the following. The set of points of the surface defined by $fghR$ in which this surface does not have normal crossings is a proper closed subset, and $E(f,g,h;R)$ is the set of one-dimensional components of this subset.

If P is any prime ideal of R, we define the incidence number $i(g,P)$ to be the number of (essentially) different prime factors of g contained in P, and we put

$i^* = \max \{i(g,P) \mid P \in E(f,g,h;R)\}$

and

$E(f,g,h;R)^* = \{P \in E(f,g,h;R) \mid i(g,P) = i^*\}$.

This means that $E(f,g,h;R)^*$ consists of those curves of $E(f,g,h;R)$ which have maximal incidence with g.

Definition. (f,g,h) is called pre-good in R if

a) $E(f,g,h;R) \neq \emptyset$.

b) $E(f,g,h;R)$ has normal crossings

c) Every $P \in E(f,g,h;R)^*$ has normal crossings with gh (i.e. $\{ghR,P\}$ has normal crossings).

T will be called good if the property of being pre-good is stable along any sequence of permissible monoidal transformations. To make this precise, we have to define permissibility and the transform of T.

Definition. A monoidal transformation $\varphi: R \to R_1$ with center P will be called permissible for (f,g,h) if $P \in E(f,g,h;R)^*$ and P has normal crossings with gh.

Definition. Let $\varphi: R \to R_1$ be a quadratic transformation or a permissible monoidal transformation for (f,g,h). Then the transform $T_1 = (f_1,g_1,h_1)$ of T will be defined as follows:

f_1 = strict transform of f

Let g_1' be a strict transform of g and $h_1' = fgh/f_1g_1'$. Then

$$g_1 = \begin{cases} g_1' & \text{if } \operatorname{ord}(R_1)(f_1) = n \\ g_1'h_1' & \text{if } \operatorname{ord}(R_1)(f_1) < n \end{cases}$$

and

$$h_1 = \begin{cases} h_1' & \text{if } \operatorname{ord}(R_1)(f_1) = n \\ 1 & \text{if } \operatorname{ord}(R_1)(f_1) < n \end{cases} \tag{7.2}$$

Note that in each case we have that $f \cdot g \cdot h = f_1 \cdot g_1 \cdot h_1$. The normal crossing condition in the definition of permissibility ensures that g_1h_1 will have normal crossings in R_1, so T_1 is in fact a tripel in the sense of (7.1). Now we can define good tripels.

<u>Definition.</u> (f,g,h) is called good in R, if the following properties hold:

a) (f,g,h) is pre-good in R.

b) Let
$$R = R_o \overset{\varphi_1}{\to} R_1 \overset{\varphi_2}{\to} \dots \overset{\varphi_t}{\to} R_t$$

be any sequence of monoidal transformations such that φ_i is permissible for $T_{i-1} = (f_{i-1}, g_{i-1}, h_{i-1})$, where $T_o = T$ and T_i is a transform of T_{i-1}. Assume that $\mathrm{ord}(R_i)(f_i) = n$ for $1 \leq i \leq t-1$. Then $\mathrm{ord}(R_t)(f_t) < n$ or T_t is pre-good in R_t.

Let us note that if $\varphi: R \to R_1$ is a permissible monoidal transformation and $T_1 = (f_1, g_1, h_1)$ a transform of T, and if (f,g,h) was pre-good in R, then T_1 need not be pre-good in R, even if $E(f_1, g_1, h_1; R_1) \neq \emptyset$. The reason is that the third condition in the definition of pre-good may be violated. As an example take

$$f = z^3 + x^7y^7$$
$$g = z + y(x^2+y^3)$$
$$h = 1$$

Then (z,y) is a permissible center, and for the corresponding transform we have

$$f_1 = z^3 + x^7y^4$$
$$g_1 = z + x^2 + y^3$$
$$h_1 = y^4$$

Now $E(f_1, g_1, h_1; R)^* = \{(z,x),(z,y)\}$ (the maximal incidence being 0), but neither (z,x) nor (z,y) has normal crossings with $g_1h_1 = zy^4 + y^4(x^2+y^3)$.

It is typical in this example that E(f,g,h;R) consists of two elements. If it consists of one element only, then $E(f_1, g_1, h_1; R_1)$ consists of at most one element (namely the exceptional divisor), therefore in this case pre-good is stable as long as $E(f_1, g_1, h_1; R_1) \neq \emptyset$. We will use this remark in §9 to prove that there are only finitely many bad points.

§8 <u>Embedded resolution - power series case</u>

Let us look at the case R = K[[x,y,z]] = S[[z]], S = K[[x,y]], again. By a suitable choice of x, y, z we may assume again that

$$f = z^n + \sum_{i=2}^{n} \alpha_i z^{n-i}, \ \alpha_i \in S, \ 2 \le i \le n \tag{8.1}$$

(note that $f = z$ if $n = 1$) and in addition

$$g = p_1 \ldots p_r, \ p_j \text{ irreducible}$$

and

$$p_j = z + \beta_j, \ \beta_j \in S, \ 1 \le j \le r. \tag{8.2}$$

Now by the definition of a transform, the process had started somewhere with $h = 1$. So by the description of the transformations of R given in §4 (under the assumption that ord(R)(f) is unchanged), we may assume that

$$h \in S. \tag{8.3}$$

Let

$$U = \{\alpha_i^{n!/i}, \ \beta_j^{n!} \mid 2 \le i \le n, \ 1 \le j \le r\}$$

and assume

$$U \text{ has normal crossings} \tag{8.4}$$

and

$$\text{For any } u,v \in U, \ uS + vS \text{ is a principal ideal.} \tag{8.5}$$

We will show that under these assumptions (f,g,h) is good in R if (and only if) f is good in R. As in §3 we will first consider the case

$$f = z^n + x^a y^b, \ a+b \ge n \quad (n > 1)^{1)} \tag{8.6}$$

where x,y is a regular system of parameters of S such that h and the β_j's are monomials in x and y.

Now since g has normal crossings, (8.2) and (8.5) imply that there are at most two prime factors p_j of g, and at most one of β_1, β_2 is in $M(S)^2$ (M(S) = the maximal ideal of S). Therefore, up to symmetry in x and y and up to units, only the following six cases are possible:

---

1) If $n = 1$, (f,g,h) is good in R if and only if $g = 1$.

Case 1. $r \leq 1$ and $g = z + x^{\lambda}$, $\lambda \geq 0$.
Case 2. $r = 1$ and $g = z + x^{\lambda}y^{\mu}$, $\lambda \geq 1$, $\mu \geq 1$.
Case 3. $r = 1$ and $g = z$.
Case 4. $r = 2$ and $p_1 = z + x$
$p_2 = z + \varepsilon x$, ε a unit $\neq 1$.
Case 5. $r = 2$ and $p_1 = z + x$
$p_2 = z + x^{\lambda}y^{\mu}$, $\lambda \geq 1$ and $\lambda+\mu \geq 2$.
Case 6. $r = 2$ and $p_1 = z + x$
$p_2 = z$.

Note that cases 3 and 6 cannot occur if $n = 1$. Assume now that f is good in R. Let us examine case 1 for $\lambda \geq 1$. By (8.5) we see that $x^{(n-1)!a}y^{(n-1)!b}S + x^{n!\lambda}S$ is principal, so $a \geq n$ and either $b = 0$ or $a \geq \lambda n$, and $E(f,g,h;R)^* = \{(z,x)\}$. Let $\varphi: R \to R_1$ be the unique permissible monoidal transformation and let (f_1,g_1,h_1) be a transform of (f,g,h). If $\text{ord}(R_1)(f_1) < n$ there is nothing to prove, so assume $\text{ord}(R_1)(f_1) = n$. Then

$$f_1 = (z/x)^n + x^{a-n}y^b \quad \text{and} \quad g_1 = z/x + x^{\lambda-1},$$

which belongs to case 1 again, and which satisfies the assumption (8.5). Similar considerations show that in each case

a) (f,g,h) is pre-good in R,
b) a permissible monoidal transformation for which $\text{ord}(R_1)(f_1) = n$ does not change the assumptions (8.4) and (8.5) and it leads to one of the cases 1-6 again.

(It may be instructive for the reader to check all these cases.) So eventually repeated permissible monoidal transformations will lead to a triple (f^*,g^*,h^*) in R^* for which

$$\begin{cases} \text{ord}_{R^*} f^* < n & \text{if } n > 1 \\ f^*g^*h^* \text{ has normal crossings} & \text{if } n = 1. \end{cases}$$

We can summarize the result of these considerations in the following way:

Let

$$R = R_o \to R_1 \to \dots \to R_t \to \dots$$

be a sequence of quadratic transformations and let $T_i = (f_i,g_i,h_i)$ be a transform of T_{i-1}, where $T_o = T$. Assume that $\text{ord}(R_i)(f_i) = n$ for all $i \geq 1$. Under the assumptions (8.4), (8.5) and (8.6) there is a t such that T_t is good in R_t.

Finally we remark that this result is true in general, since assumptions (8.4) and (8.5) will be satisfied in R_i for large i, and the general case (8.1) can be reduced to (8.6) as in §4.

§9 Bad points for embedded resolution are finite in number

Let Z be a regular, 3-dimensional, excellent scheme and let (F,G,H) be a triple of surfaces on Z such that

F and G ∪ H have no common components,
G ∪ H has normal crossings, (9.1)
F is reduced.

Let P be a point on F, let R be the local ring of Z at P and let F, G, H be defined in R by f, g, h respectively.
Definition. P will be called a good (pre-good) point for (F,G,H) if $\mathrm{mult}_P(F) < m(F)$ or the tripel (f,g,h) is good (pre-good) in R.

Now let C be a curve on F.
Definition. C will be called permissible for (F,G,H) if
a) C is regular;
b) $C \subset M(F)$
c) among all curves on F satisfying a) and b), C is contained in the maximal possible number of components of G;
d) C has normal crossings with G ∪ H.

A monoidal transformation $\varphi: Z_1 \to Z$ with center C will be called permissible for (F,G,H) if C is permissible for (F,G,H). Note that a permissible curve is permissible at each of its points in the sense of the local definition of §7.

Let $\varphi: Z_1 \to Z$ be a quadratic transformation centered at a point $P \in M(F)$ or a permissible monoidal transformation for (F,G,H). Let F_1 (resp. G_1) be a strict transform of F (resp. G) under φ and let $H_1 = \varphi^{-1}(H \cup C)$, where C is the center of φ. Then the transform of (F,G,H) is defined to be

$$(F_1, G_1, H_1) \quad \text{if } m(F_1) = m(F)$$
$$(F_1, G_1 \cup H_1, \emptyset) \quad \text{if } m(F_1) < m(F).$$

As noted in the local part, the transform of (F,G,H) will again be a triple satisfying (9.1). Note that in both cases $\varphi^{-1}(F \cup G \cup H) = F_1 \cup G_1 \cup H_1$. To prove the finiteness of bad points for (F,G,H) we proceed as in §5.
We define

$$M(F,G,H) = \begin{cases} M(F) \text{ if } m(F) > 1 \\ \{P \in F \mid F \cup G \cup H \text{ does not have normal crossings at } P\} \text{ if } m(F) = 1 \end{cases}$$

and we define B(F,G,H) to consist of the isolated points, the singularities of M(F,G,H), and those regular points of M(F,G,H), at which $G \cup H$ does not have normal crossings with M(F,G,H). (Note that M(F,G,H) is a proper closed subset of F also in the case m(F) = 1). If we can show that B(F,G,H) has properties analogous to (5.1), then the argument used in §5 will show that there are only finitely many points which are bad (= not good) for (F,G,H). a) and b) are clear, and so is c) in case m(F) > 1. To obtain c) in case m(F) = 1 we use again the fact that permissible monoidal transformations correspond to quadratic transformations of plane curves. Then the translation of c) into local algebra simply means that along a quadratic sequence of 2-dimensional regular local (excellent) rings any finite set of elements will have normal crossings eventually.

§10 Review of the procedure of embedded resolutions

To describe an intermediate step of embedded resolution, it is convenient to start with a pair (F,G) of surfaces rather than F only. Using the notation of §9 we assume $F,G \subset Z$, F and G have no common components and G has normal crossings. Therefore $(F,G,\emptyset)$ is a triple in the sense of (9.1). According to §9 there are only finitely many bad points for $(F,G,\emptyset)$, and by §8 they can be removed by finitely many quadratic transformations. If m(F) > 1, a finite number of permissible monoidal transformations will transform (F,G,H) into a triple $(F^*,G^*,\emptyset)$ with $m(F^*) < m(F)$. If m(F) = 1, the same procedure will lead to a triple (F^*,G^*,H^*) such that $F^* \cup G^* \cup H^*$ has normal crossings, and this proves theorem B.

IV Auxiliary results for surfaces

In this chapter, R denotes a three-dimensional, excellent, regular local ring with maximal M and residue field K. We assume that K has characteristic 0. (This assumption will only be used in §3 and §4.) Furthermore we fix a nonzero element f in R such that $\mathrm{ord}(R)(f) = n > 0$.

§1 Equimultiple curves under quadratic transformations

Let $R \to R_1$ be a quadratic transform of R, let f_1 a strict transform of f in R_1 and assume $\mathrm{ord}(R_1)(f_1) = n$.

Let $P_0 \in E(f_1,R_1)$ and assume first that $P_0 \cap R = M$. By the correspondence between R_1/MR_1 and $\mathrm{gr}(R)$ described in I, §3, P_0 corresponds to a homogeneous prime ideal Q of $\mathrm{gr}(R)$, necessarily of height one, such that $\mathrm{in}(f) \in Q^{(n)}$. As observed earlier, these conditions imply that Q is generated by $\mathrm{in}(z)$ for some $z \in M \setminus M^2$, and $\mathrm{in}(f) = \mathrm{in}(z)^n$ up to a unit. It follows that $P_0 = (x,z/x)$, where $x \in R$ is chosen so that $MR_1 = xR_1$. Furthermore P_0 is unique and R_1/P_0 is regular.

If $P_1 \in E(f_1,R_1)$ and $P_1 \cap R \neq M$, then

$$(R_1)_{P_1} = R_P \quad \text{where } P = P_1 \cap R.$$

Since $f_1 \cdot R_P = f \cdot R_P$, $P \in E(f,R)$. So we obtain

$$E(f_1,R_1) = \begin{cases} \{PR_P \cap R_1 \mid P \in E(f,R) \text{ and } R_P \supset R_1\} & \text{or} \\ \{PR_P \cap R_1 \mid P \in E(f,R) \text{ and } R_P \supset R_1\} \cup \{P_0\}. & \end{cases}$$

Proposition 1. Let $R \to R_1$, be a quadratic transform and let f_1 be a strict transform of f in R_1. Assume that $\mathrm{ord}(R_1)(f_1) = n$ and that $E(f,R)$ has normal crossings. Then $E(f_1,R_1)$ has normal crossings.

Proof. Let x,y,z be a regular system of parameters of R whose subsets generate the elements of $E(f,R)$, and assume $MR_1 = xR_1$. If $P_1 \in E(f_1,R_1)$, then either $P_1 \cap R = M$ or $P_1 \cap R = (y,z)$, in which case we must have $P_1 = (\frac{y}{x},\frac{z}{x})$. Therefore if $E(f_1,R_1)$ has only one element, then there is nothing to prove.

Assume therefore that $E(f_1,R_1) = \{P_0,P_1\}$, where $P_0 \cap R = M$, $P_1 = (y/x)R_1+(z/x)R_1$. Since x, y/x, z/x is a regular system of parameters of R_1 and R_1/P_0 is regular, we have $P_0 = xR_1 + (a(y/x) + b(z/x))R_1$, where $a, b \in R_1$ and either a or b is a unit in R_1. If e.g. $a = 1$, then we write $P_1 = ((y/x) / b(z/x))R_1 + (z/x)R_1$ to see that $E(f_1,R_1)$ has normal crossings.

§2 Equimultiple curves under monoidal transformations

Let $R \to R_1$ be a monoidal transform with center $P \in E(f,R)$ (remember that monoidal includes the assumption that R/P is regular). Let f_1 be a strict transform of f in R_1 and assume $\mathrm{ord}(R_1)(f_1) = n$. By the considerations made in I,§4, we can find a regular system of parameters x, y, z of R with the following properties:

$$\left.\begin{array}{l} P = xR + yR,\ PR_1 = xR_1 \\ x,\ y/x,\ z \text{ is a regular system of parameters of } R_1 \\ f-ay^n \in P^{n+1} + zP^n \text{ for some unit a in } R. \end{array}\right\} \qquad (2.1)$$

Assume that P_1 is a prime ideal in R_1 such that $f_1 \in P_1$ and $P_1 \cap R \supset P$. If $P_1 \cap R = M$, we must have $P_1 = xR_1 + zR_1$. From (2.1) we see that $f_1-a(y/x)^n \in P_1$, which leads to the contradiction $y/x \in P_1$. Therefore $P_1 \cap R = P$.

Applying this remark to an element $P_1 \in E(f_1,R_1)$ such that $P_1 \cap R \supset P$ we see that $(R_1)P_1$ dominates R_P, and actually $(R_1)_{P_1}$ is the unique quadratic transform of R_P such that $\mathrm{ord}((R_1)_{P_1})(f_1) = n$. On the other hand, if $P_1 \in E(f_1,R_1)$ such that $P_1 \cap R \not\supset P$, then we conclude as in the preceding section that $P_1 \cap R \in E(f,R)$.

Proposition 2. Let $R \to R_1$ be a monoidal transform of R with center $P \in E(f,R)$, f_1 a strict transform of f in R_1. Assume that $\mathrm{ord}(R_1)(f_1) = n$ and that $E(f,R)$ has normal crossings. Then $E(f_1,R_1)$ has normal crossings.

Proof. Choose a regular system of parameters x, y, z whose subsets generate the elements of $E(f,R)$, and assume $P = xR + yR$ and $PR_1 = xR_1$. If $E(f_1,R_1) = \{P_1\}$ and $x \in P_1$, there is nothing to prove. If $P_2 \in E(f_1,R_1)$ such that $x \notin P_2$, then $P_2 \cap R = yR + zR$ and consequently $P_2 = (y/x)R_1 + zR_1$, so we are done if $E(f_1,R_1)$ consists of one element only. Therefore we assume now that $E(f_1,R_1) = \{P_1,P_2\}$ and $x \in P_1$, $P_2 = (y/x)R_1 + zR_1$. Since x, y/x, z is a regular system of parameters of R_1, and since R_1/P_1 is regular, there are elements $b,c \in R_1$ such that P_1 is generated by x and $b(y/x) + cz$, and either b or c is a unit in R_1. As in Proposition 1 we see that $\{P_1,P_2\}$ has normal crossings.

§3 Equimultiple curves under completion

Let $f \in R$ be a nonzero element such that $\operatorname{ord}(R)(f) = n > 0$ and assume that fR is not an nth power of any principal ideal of R.

Lemma 1. If $P^* \in E(f,R^*)$ then $P^* \cap R \in E(f,R)$.
Proof. Let $P = P^* \cap R$. Write

$$fR = p_1^{n_1} \cdot \ldots \cdot p_s^{n_s} R,$$

where $p_1,\ldots,p_s$ are the distinct prime factors of f in R. Then $f \in P$ implies $p_i \in P$ for some i, say $p_1 \in P$. Let us show first that $\operatorname{ht}(P) = 2$. If not then $P = p_1R$ and we conclude that

$$f(R^*)_{P^*} = p_1^{n_1}(R^*)_{P^*}.$$

Since R is excellent, $R^*_{P^*}/p_1R^*_{P^*}$ is regular, necessarily of dimension one, so $\operatorname{ord}(R^*_{P^*})(p_1) = 1$ and we obtain

$$n_1 = n, \ \operatorname{ord}(R)(p_1) = 1$$

so $fR = p_1^nR$. Since this was excluded by assumption, we must have $\operatorname{ht}(P) = 2$. In particular P^* is minimal among the primes of R^* containing PR^*. Using excellence again we see that $R^*_{P^*}/PR^*_{P^*}$ is regular of dimension zero and therefore $PR^*_{P^*} = P^*R^*_{P^*}$. Since $R_P \to R^*_{P^*}$ is faithfully flat, we have

$$M(R^*_{P^*})^m \cap R_P = P^mR^*_{P^*} \cap R_P = P^mR_P \quad \text{for all } m$$

so that $f \in P^nR_P$; i.e. $P \in E(f,R)$.

Using the notation of the proof above, we note that if R/P is regular, then R^*/P^* is regular and P^* the unique element of $E(f,R^*)$ which contains P. From this remark and Lemma 1 we obtain immediately

Lemma 2. If $E(f,R)$ has normal crossings, then $E(f,R^*)$ has normal crossings. Therefore f pre-good in R implies f pre-good in R^*.

The converse of Lemma 2 need not be true. As an example, take $R = K[x,y,z]_{(x,y,z)}$ and

$$f = z^2 + [x^2 - y^2 + y^3]^2.$$

Then $R^* = K[[x,y,z]]$, and since $x^2-y^2 = (x+y)(x-y)$, $x^2-y^2+y^3$ can be factored in R^* by Hensel's Lemma:

$$x^2 - y^2 + y^3 = g_1\cdot(x-y)\cdot g_2\cdot(x+y), \quad g_1, g_2 \in R^*.$$

Therefore $E(f,R^*) = \{(z,x-y),(z,x+y)\}$ but $E(f,R) = \{(z,x^2-y^2+y^3)\}$, so $E(f,R^*)$ has normal crossings, but $E(f,R)$ does not.

Let $P \in E(f,R)$ and let P^* be a minimal prime ideal of R^* containing PR^*. Then certainly $P^* \in E(f,R^*)$. If now R^*/P^* is regular but R/P is not, then there must be another prime ideal Q^* of R^* such that $Q^* \cap R = P$. (Note that PR^* is an intersection of prime ideals since R is excellent.) This shows that the example given above is typical in the following sense: If $E(f,R^*)$ has normal crossings and $E(f,R)$ does not, then there are $P^*, Q^* \in E(f,R^*)$ such that $P^* \neq Q^*$ and $P^* \cap R = Q^* \cap R$. In particular $R/P^*\cap R$ is not regular.

Therefore if $E(f,R)$ has normal crossings, then the elements of $E(f,R)$ and $E(f,R^*)$ correspond uniquely to each other by extension and contraction. Taking into account that any monoidal transformation of R induces uniquely a monoidal transformation of R^*, we have

Lemma 3. If $E(f,R)$ has normal crossings, then f is good in R if and only if f is good in R^*.

In II, §4, we have shown the following

Proposition 3*. Let

$$R^* = R_0^* \to R_1^* \to \dots \to R_i^* \to \dots$$

be a sequence of three-dimensional regular local rings such that each R_i^* is the completion of a quadratic transform of R_{i-1}. Let $f = f_0 \in R^*$ such that $f \neq 0$ and $\mathrm{ord}(R^*)(f) = n > 0$. Define $f_i \in R_i^*$ inductively to be a strict transform of f_{i-1}, and assume that $\mathrm{ord}(R_i^*)(f_i) = n$ for all i. Then for some j, f_j is good in R_j^*.

We can use this now to prove

Proposition 3. Let

$$R = R_o \to R_1 \to \ldots \to R_i \to \ldots$$

be a sequence of excellent three-dimensional regular local rings such that each R_i is a quadratic transform of R_{i-1}. Let $f = f_o \in R$ such that $f \neq 0$ and $\mathrm{ord}(R)(f) = n > 0$. Define $f_i \in R_i$ inductively to be a strict transform of f_{i-1}, and assume that $\mathrm{ord}(R_i)(f_i) = n$ for all i. Then for some j, f_j is good in R_j.

Proof. By Proposition 3* we may assume f good in R*. Then by Lemma 3, if f is not good in R, E(f,R) does not have normal crossings. Now $E(f_1,R_1^*)$ has normal crossings by Proposition 1, and it contains at most one element P_1^* such that $P_1^* \not\supset M(R^*)$ and at most one element P_2^* such that $P_2^* \supset M(R^*)$. It follows that $E(f_1,R_1)$ has normal crossings, and by Proposition 1, $E(f_i,R_i)$ has normal crossings for all $i \geq 1$. Of course f_1 may not be good in R_1^* (as shown by the example in I, §2). But by Proposition 3*, f_j is good in R_j^* for some $j \geq 1$. It follows from Lemma 3 that f_j is good in R_j.

Let (f,g,h) be a triple in R in the sense of II, (7.1). It is clear that corresponding to Lemma 3 we have

Lemma 4. If E(f,R) has normal crossings, then (f,g,h) is good in R if and only if (f,g,h) is good in R*.

Therefore by the same argument as in Proposition 3, the result of II, §8, yields

Proposition 4. Let

$$R = R_o \to R_1 \to R_2 \to \ldots \to R_i \to \ldots$$

be a sequence of excellent three-dimensional regular local rings such that each R_i is a quadratic transform of R_{i-1}. Let $(f,g,h) = (f_o,g_o,h_o)$ be a triple in R and define (f_i,g_i,h_i) inductively to be the triple of

R_i which is a transform of $(f_{i-1},g_{i-1},h_{i-1})$. Assume that $\mathrm{ord}(R_i)(f_i) = n$ for all $i \geq 0$. Then for some j, (f_j,g_j,h_j) is good in R_j.

§4 Use of the Tschirnhausen transformation

In this section we take R to be the power series ring $K[[x,y,z]]$ over a field K of characteristic zero. Let $f \in R$, $f \neq 0$, and $n = \mathrm{ord}(R)(f)$. By the Weierstraß Preparation Theorem and related results we may choose x,y,z in such a way that the ideal generated by f is the same as the ideal generated by $z^n + \sum_{i=1}^{n} a_i z^{i-n}$, where $a_i \in K[[x,y]]$. The Tschirnhausen transformation consists in replacing z by $z^* = z-a_1/n$, so f becomes a polynomial in z^* in which the coefficient of $(z^*)^{n-1}$ is zero. So from now on we assume

$$f = z^n + \sum_{i=2}^{n} a_i z^{n-i}, \quad a_i \in S = K[[x,y]], \ n > 0.$$

Let $R \to R_1$ be a quadratic transform and f_1 a strict transform of f in R_1, and assume $\mathrm{ord}(R_1)(f_1) > 0$. We claim that

$$M(R)R_1 = \begin{cases} xR_1 & \text{or} \\ yR_1. & \end{cases}$$

Assume the contrary. Then x/z and y/z are in $M(R_1)$ and therefore

$$f_1 = f/z^n = 1 + \sum_{i=2}^{n} a_i/z^i$$

is a unit in R_1. (Note that $\mathrm{ord}(R)(f) = n$ implies $\mathrm{ord}(S)(a_i) \geq i$ and therefore $a_i/z^i \in M(R_1)$. Note also that this argument is valid without assuming $a_1 = 0$.)

Assume in addition that $\mathrm{ord}(R_1)(f_1) = n$, and let $M(R)R_1 = xR_1$. Let Q be the homogeneous prime ideal of $\mathrm{gr}(R)$ such that $R_1/xR_1 = \mathrm{gr}(R)_{(Q)}$. Then $\mathrm{in}(f) \in Q^{(n)}$. Now

$$\mathrm{in}(f) = \mathrm{in}(z)^n + \sum_{i=2}^{n} \bar{a}_i \ \mathrm{in}(z)^{n-i}$$

where $\bar{a}_i$ is the class of a_i mod $M(R)^{i+1}$. Using derivatives as in II, §4, we see that $\mathrm{in}(z) \in Q$, which means that $z/x \in M(R_1)$ and R_1 has as regular system of parameters x, z/x, $y^* = F(y/x)$, where $F(T) \in S$ (see I, §4). It follows that y^* defines a quadratic transform S_1 of S and

$$R_1^* = S_1^*[[z/x]]$$

We turn now to a monoidal transform $R \to R_1$ with center $P \in E(f,R)$. As usual, f_1 denotes a strict transform of f in R_1. By the observation made in II, §4, we have

$$P = zR + tR,\ t \in S \text{ and } t^i | a_i \quad \text{for all } i.$$

Since $\mathrm{ord}(S)(t) = 1$, we may assume $t = x$. As in the case of a quadratic transform one sees that $\mathrm{ord}(R_1)(f_1) > 0$ implies $PR_1 = xR_1$. If we assume $\mathrm{ord}(R_1)(f_1) = n$, then we have

$$R_1/xR_1+yR_1 \simeq (gr_P(R) \otimes {}_R K)_{(Q)}$$

for some homogeneous prime ideal Q, and the residue class of $in_P(f)$ belongs to $Q^{(n)} = Q^n$. As above we conclude that $z/x \in M(R_1)$, and x, y, z/x is a regular system of parameters of R_1. It follows that

$$R_1^* = S[[z/x]].$$

References

[1] Abhyankar S.S. (1966) Resolution of singularities of embedded algebraic surfaces. Academic Press, New York London

[2] Hironaka H. (1964) Resolution of singularities of an algebraic variety over a field of characteristic zero I, II. Annals of Mathematics 79:109-326

[3] Lipman J. (1975) Introduction to resolution of singularities, in Algebraic Geometry Arcata 1974. Amer. Math. Soc. Proc. Symp. Pure Math. 29:187-230

[4] Matsumura H. (1970) Commutative algebra. Benjamin, New York

[5] Serre J-P. (1965) Algèbre Locale - Multiplicités. Springer Lecture Notes in Mathematics 11, Berlin Heidelberg New York

[6] Zariski O. (1967) Exceptional singularities of an algebroid surface and their reduction, Accad. Naz. Lincei Rend. Cl. Sci. Fis. Mat. Natur. Serie VIII, 43:135-146

[7] Zariski O. (1978) A new proof of the total embedded resolution theorem for algebraic surfaces (based on the theory of quasi-ordinary singularities). Amer. J. Math. 100:411-442

[8] Abhyankar S.S. (1982) Weighted Expansions for Canonical Desingularization. Springer Lecture Notes in Mathematics vol. 910

DESINGULARIZATION IN LOW DIMENSION

Jean GIRAUD

§ 1. Smart and coarse theorems.

There are two problems, namely desingularization and simplification of a boundary and for each of them two levels of precision in the corresponding theorems wich I will call coarse and smart.

Coarse desingularization theorem. Let X be a scheme (or a complex analytic space). Assume that X is reduced. There exists a proper morphism $\pi: X' \longrightarrow X$, such that the induced morphism $\pi^{-1}(X_{reg}) \longrightarrow X_{reg}$ is an isomorphism, where X_{reg} is the set of points where the local ring $O_{X,x}$ is regular, and X' is regular.

This is a theorem in the complex analytic case or for X excellent of characteristic zero or for X excellent and $\dim X \leq 2$.

Coarse Simplification of boundary. Let Z be a regular scheme (or smooth complex analytic space), let Y be a closed subset of Z. There exists a proper and birational morphism $\pi: Z' \longrightarrow Z$ such that

(1) Z' is regular

(2) $\pi^{-1}(Y)$ is a normal crossing divisor,

(for short we say that $\pi^{-1}(Y)$ is a d.n.c.) and each irreducible of $\pi^{-1}(Y)$ is regular.

Of course if Y is regular we achieve that by blowing up Y.

There are various smart versions of these theorems and it is not my intention to discuss them. I will only give one of them.

Smart desingularization theorem. Let X be an excellent scheme of characteristic zero.

There exists a sequence

$$\begin{array}{ccccccc} X = X_0 & \longleftarrow & X_1 & \longleftarrow & X_2 \ \ldots\ldots & X_{N-1} & \longleftarrow & X_N \\ \cup & & \cup & & \cup & \cup & & \\ Y_1 & & Y_2 & & Y_3 & Y_N & & \end{array}$$

such that

(i) Y_{i+1} is regular, closed in X_i, and contained in the singular locus of X_i, $0 \leq i \leq N-1$.

(ii) X_{i+1} is the blowing up of X_i with center Y_{i+1}

(iii) X_N is regular

(iV) if E_i is the inverse image of Y_i in X_i, $1 \leq i \leq N$, then $\bigcup_{1 \leq i \leq N} E_i$ is a divisor with normal crossings.

Here the important condition is (i), because if you start with some embedding of X as a closed subscheme of a regular scheme Z, then by letting $Z_0 = Z$, and Z_{i+1} = blowing up of Z_i with center Y_{i+1} you get that each X_i is closed in a regular Z_i. In other words if X_i is embedded in a regular Z, then the desingularization X_N is embedded in a regular Z_N.

Remark 1. If X is embedded as a hypersurface in a regular Z, then the coarse simplification of boundary will produce $\pi: Z' \longrightarrow Z$ such that $\pi^{-1}(X)$ is a d.n.c. Hence, if X is reduced and irreducible we will get a birationnal desingularization $X' \longrightarrow X$, by taking for X' the component of $\pi^{-1}(X)$ whose image is a divisor in Z. This will only be a coarse desingularization of X, unless we have a smart simplification of boundary, by wich I mean some condition implying condition (i) of the smart desingularization theorem.

Remark 2. On the other hand if X is a curve in $\mathbb{P}_3 = Z$, then simplification of boundary will not give a desingularization of X since the strict transform of X is going to be empty because one will have to blow up X if one wants $\pi^{-1}(X)$ to be a Cartier divisor: this is the universal property of blowing up.

Hence desingularization and simplification of boundary are not at all the same problem. They are closely related in the hypersurface case and in that case some people call it "embedded resolution". In the general case desingularization and simplification are proved simultaneously each one being useful for the proof of the other one but not in such a naive way as is suggested by the hypersurface case.

To end this paragraph I would like to point out that to prove one of the three statements in dimension N then you need as an induction hypothesis a very smart statement in dimension N-1 . This is not exactly true in low dimension, for instance use will see a nice proof of coarse desingularization of X , if dim(X) = 2 due to J.Lipman. And also simplification of a boundary Y in Z with dim(Z)=2 , is automatically smart since one only has to blow up points. In some sense the case of surfaces is somewhat misleading and specially the surface-hypersurface case (See § 4 and 5).

§ 2. What can be achieved by blowing up closed points.

If dim Z = 1 and Z is regular , then any closed subscheme is a divisor with normal crossing: nothing to prove.

If dim X = 1 , and X excellent and reduced, then X_{sing} is a discrete set of closed points; by blowing them up one gets $X_1 \longrightarrow X$ which is different from X because for $\xi \in X_{sing}, m_{X,\xi}$ is not principal, hence the blowing up of ξ gives $X' \neq X$. Repeating this operation one must reach the normalization of X since this normalization is finite over X . Observe that for curves there is a unique coarse desingularization, namely the normalization, which is also, as we just saw, a smart desingularization.

If dim $Z = 2$ and Z is regular, if J is a coherent sheaf of ideals of Z, then the set of $\xi \in Z$ such that $J\, O_{Z,\xi}$ is not a d.n.c. is a discrete set of closed points; call it F. Blowing up of F in Z gives Z_1 which is regular and we get a new exceptionnal set F_1 for $J_1 = JO_{Z_1}$, and so on. One can prove that by repeating this process one gets $F_n = \phi$. In other words simplification of boundary is automatic in dimension 2, and is achieved by repeated blowing up of closed points.

Some isolated singularities of surfaces can be solved by (repeated) blowing up of closed points. The most obvious example is the cone over a projective smooth curve C since blowing up the vertex of the cone gives the line bundle on C given by the inclusion $C \subset \mathbb{P}^N$. Less obvious is the case of a normal surface X which admits a finite and flat projection over a regular surface, such that the discriminant has normal crossings, (car.o) . This is the basis of JUNG'S method. **Another** example is given by normal surfaces with rational singularities: this is the basis of LIPMAN'S proof of coarse desingularization theorem for excellent schemes of dimension two.

But it is easy to give an example of a normal surface such that the blowing up of the closed point gives a surface with non isolated singularity: $z^3 = x^5 + y^5$. The blowing up is covered by two pieces. The first one has coordinates $z' = z/x, x, y' = y/x$ and equation $z'^3 = x'^2(1+y'^5)$ and the second one is obtained by exchanging x and y.

§ 3. Coarse desingularization of surfaces.

Since we just saw that blowing up of points will not desingularize surfaces, the next idea is to perform sucessively blowing up of closed points (the singular locus) and normalization.Due to ZARISKI'S and ABHYANKAR'S work we know that this procedure works but the simplest proof has been given recently by LIPMAN. Let us state his result.

Theorem Let X be an excellent scheme of dimension two. Assume that X is normal and quasi compact. Consider a sequence

$$\begin{array}{ccccccccc} X = X_0 & \longleftarrow & X_1 & \longleftarrow & X_2 & \cdots\cdots X_n & \longleftarrow & X_{n+1} & \cdots\cdots \\ \cup & & \cup & & \cup & \cup & & \cup & \\ F_1 & & F_2 & & F_3 & F_{n+1} & & F_{n+2} & \end{array}$$

Such that

(i) F_{i+1} is a discrete subset of $\mathrm{Sing}(X_i)$, $i > o$

(ii) X_{i+1} is the normalization of the blowing up of X_i with center F_{i+1} , $i > o$

Then

(A) there exists an integer N such that for $n > N$ the inclusion $H^1(X_N, O_{X_N}) \subset H^1(X_n, O_{X_n})$ is bijective

(B) if $F_{i+1} = \mathrm{Sing}\,(X_i)$ for any i , then there exists an N such that F_{N+1} is empty, hence X_N is regular.

The largest part of LIPMAN'S paper is devoted to the proof of (A). This is due to the fact that he wants to give a self contained account of local duality. If you accept it you know that there is an injective map $f_{n*}(\omega_{X_n}) \longrightarrow \omega_{X_0}$, where $f_n : X_n \longrightarrow X_0$ is the projection and ω is the dualizing sheaf. Grauert-Riemenscheider vanishing theorem says that $\mathbb{R}^1 f_{n*}(\omega_{X_n}) = 0$ provided that X admits a desingularization, but since he does not know yet the existence of a desingularization, LIPMAN has to prove this vanishing theorem. When this is done local duality tells you that the length of $H^1(X_n, O_{X_n}) / H^1(X_0, O_{X_0})$ is the length of $\omega_{X_0} / f_{n*}(\omega_{X_n})$. Now one needs some kind of <u>discriminant</u> or trace map to show that there exist a $d \in \omega_X(X)$ (Say X is affine) such that $d\omega_{X_0} \subset f_{n*}(\omega_{X_n})$ <u>for any</u> n .

The next step is to see that one only has to study points of X_n which are on the strict transform of the curve $d = o$ and for these points O_{X_n, x_n} is regular for n big enough: this is a classical easy remark.

This is essentially the argument for (A), see LIPMAN'S paper for details!

Now to prove (B) it is enough to show that for a sequence of normalized blowing up as in (B), such that $H^1(X_o, O_{X_o}) = H^1(X_n, O_{X_n})$ for all $n \geq 0$, X_n is regular for n big enough. First there is an older lemma of LIPMAN saying that in that case the blowing up remains normal: in other words you don't have to normalize. Second the singularity has very special features (all extracted from the cohomological hypothesis) wich make the proof rather easy, at least for somebody who has some familiarity with the manipulation of infinitely near points.

I apologize for all the imprecisions contained in this short description of a rather long and very elegant paper.

The oldest idea for proving the coarse desingularization theorem is due to JUNG and is 100 years old. It is still as fresh and beautiful as on the first day: it only works in characteristic 0.

Say that X is a <u>projective surface</u> and look for a finite morphisme $\pi : X \longrightarrow S = \mathbb{P}_2$ given by some <u>generic linear projection</u>. Look for the discriminant Δ of that projection: if X is not a hypersurface one takes for Δ a suitable Fitting ideal of the relative differentials $\Omega^1_{X/S}$. This Fitting ideal is equal to the usual discriminant in the hypersurface case; see for instance TEISSIER, Arcata 1975. As we saw in § 2, a suitable sequence of blowing up of closed points in S will give $p : S' \longrightarrow S$ such that $\Delta' = p^{-1}(\Delta)$ is a divisor with normal crossings. Let X' be the pull back of X by p and let $\overline{X'}$ be its normalization. The claim is that $\overline{X'}$ can be desingularized by repeated blowing up of closed points (one does not have to normalize again). Before proving this we observe that if it is true we get $X'' \longrightarrow \overline{X'} \longrightarrow X$, which is birational, with X'' regular, but we may have modified X above some points $\xi \in X$ such that $O_{X,\xi}$ is regular but Δ is not normal crossings at $\pi(\xi) \in S$. In other words we cannot say that X'' is a desingularization of X, because $X'' \longrightarrow X$ is not an isomorphism above X_{reg} (see the first statement page 1). This is not difficult to arrange. In fact take an open neighbourhood U of ξ in X such that $U \cap \pi^{-1}(\mathrm{Sing}\,\Delta) = \xi$. Such a U exists since $\mathrm{Sing}\,(\Delta)$ is a finite set and $\pi: X \longrightarrow S$ is finite. Then patch U and $X''-q^{-1}(\xi)$ along $U-\{\xi\}$, where $q: X'' \longrightarrow X$ is the projection. We have contracted $q^{-1}(\xi)$ to ξ. If we do it for all the $\xi \in \pi^{-1}(\mathrm{Sing}\,\Delta)$ such that $O_{X,\xi}$ is regular, we get a coarse desingularization of X.

Now we go to the heart of JUNG's idea.

Definition Let X be a normal surface. We say that X can be desingularized by blowing up of points if there exist a finite sequence

$$\begin{array}{ccccccccc} X = X_0 & \longleftarrow & X_1 & \longleftarrow & X_2 & \cdots\cdots\cdots & X_{N-1} & \longleftarrow & X_N \\ \cup & & \cup & & \cup & & \cup & & \cup \\ F_1 & & F_2 & & F_3 & & F_N & & F_{N+1} \end{array}$$

such that (i) X_i is normal for $o < i < N$

(ii) F_{i+1} is the singular locus of X_i for $0 < i < N$

(iii) X_{i+1} is obtained by blowing up of X_i with center F_{i+1}, $1 < i < N$.

(iv) F_{N+1} is empty.

Proposition: Let S be a regular algebraic surface over $\mathbb{C}$. Let $X \longrightarrow S$ be a finite covering. Assume that X is normal. Let Δ be a normal crossing divisor of S such that if we let $S_0 = S-\Delta$ and if $\pi_0 = X_0 \longrightarrow S_0$ is the restriction of π to S_0 then π_0 is etale. Then X can be desingularized by blowing up of points.

With this formulation it is obvious that we can replace X and S by the corresponding analytic spaces: in fact blowing up of closed points commutes to that operation. Hence we can assume that we are in the complex analytic case and that $S = D \times D$ and $S_o = D' \times D'$, where D is the unit disk of the complex line $\underline{\underline{C}}$ and $D' = D - \{o\}$.

Let $X_{o,\alpha}$ be a connected component of X_o. It is a finite etale covering of S_o, hence $X_{o,\alpha}$ is given by a finite quotient group Q of $\pi_1(S_o) = \mathbb{Z} \times \mathbb{Z}$. Let $G = \mathrm{Ker}(\mathbb{Z} \times \mathbb{Z} \longrightarrow Q)$. If $G = a\,\mathbb{Z} \times b\mathbb{Z}$ with integers a and b, then the covering $X_{o,\alpha}$ is isomorphic to $D' \times D' \longrightarrow D' \times D'$, $(x,y) \longmapsto (x^a, y^b)$. Furthermore such a covering can be extended to a ramified covering of S by a <u>smooth</u> space namely $D \times D \longrightarrow D \times D = S$, $(x,y) \longmapsto (x^a, y^b)$. Of course G need not be such a diagonal subgroup. Nevertheless there exists a diagonal subgroup G_1 of $\mathbb{Z} \times \mathbb{Z}$ such that $G_1 \subset G \subset \mathbb{Z} \times \mathbb{Z}$ and such that G/G_1 is cyclic. This is an exercise about integral lattices whose solution gives the following. Let (e_1, e_2) be the canonical basis of $\mathbb{Z} \times \mathbb{Z}$. There exist integers $a,b,c,$ such that $G = \mathbb{Z}\, ae_1 + \mathbb{Z}\,(be_1 + de_2)$. Let $\alpha = \mathrm{G.C.D.}(a,b)$ and $a' = a/\alpha$ Let $G_1 = \mathbb{Z}\, ae_1 + \mathbb{Z}\, a'd\, e_2$ then $G_1 \subset G$, and G/G_1 is generated by $\xi_2 = be_1 + de_2$, with $a'\xi_2 \in G_1$. The ramified covering corresponding to G_1 is $D \times D \longrightarrow D \times D$, $(x,y) \longmapsto (x^a, y^{a'd})$, the Galois group $Q_1 = (\mathbb{Z} \times \mathbb{Z})/G_1$ acts via two primitive roots of unity ξ, η with $\xi^a = \eta^{a'd} = 1$. The generator ξ_2 of the subgroup G/G_1 acts by $(x,y) \longmapsto (\xi^b x, \eta^d y)$ which can be written as $(x,y) \longmapsto (t^p x, ty)$ with $t = \eta^d$, with suitable p and $t^{a'} = 1$, $(a',p) = 1$. Hence the quotient of $D \times D$ by G/G_1 is the well known cyclic singularity labelled $(a',1,p)$. It is a toroïdal embedding (in French it is called an éventail). Now I claim that this éventail is nothing but the closure in X of the component $X_{o,\alpha}$ we started with. In fact one has a diagram

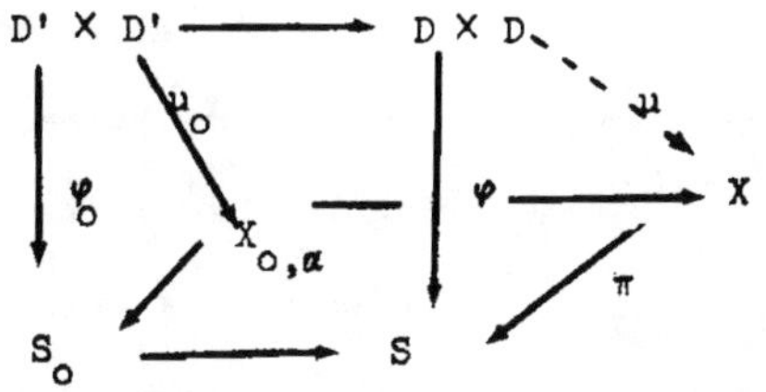

where $\varphi(x,y)=(x^{a},y^{a'd})$, and where u_0 exists since $G_1 \subset G$. The morphism u_0 can be extended to $D \times D$ (dotted arrow) because $D \times D$ is smooth and X is finite over S . Since u_0 gives an isomorphism $D'XD'/(G/G_1) \xrightarrow{\sim} X_{0,\alpha}$, the morphism u factors through the quotient $X_\alpha = D \times D/(G/G_1)$ and we get

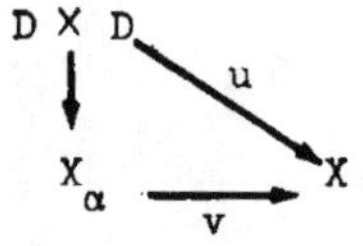

Since X_α and X are finite over S the image of v is the closure $\overline{X_{0,\alpha}}$ of $X_{0,\alpha}$ in X . Since X is normal and $X_{0,\alpha}$ is a connected component of X_0 one has that $\overline{X_{0,\alpha}}$ is a connected component of X . Now $\overline{X_{0,\alpha}}$ and X are both normal and v is an isomorphism when restricted to S_0 ; hence v is an isomorphism because v is finite.

Eventually we get that X is a finite union of toroïdal embeddings. By explicit computation one knows that a toroïdal embedding can be desingularized by repeated blowing up of closed points. This finishes the proof of the proposition.

In LAUFER's book, or HIRZEBRUCH's thesis one can find an explicit description of the minimal resolution of X in terms of continued fractions. This is a special feature of toroïdal embeddings of dimension 2 and a very useful one too. But if one follows this line of argument it is not so easy to deduce from it that <u>algebraic</u> surfaces can be desingularized in the sense that the desingularization is also an algebraic variety: in other words X is only analytically isomorphic to a toroïdal embedding. It is to avoid this kind of difficulty that I choosed to talk in terms of desingularization by blowing up of points, a property which obviously goes back from analytic to algebraic geometry.

An old dream of many specialists is to extend JUNG's method to dimension 3. We first notice that we would need simplification of the boundary Δ inside S with $\dim S = 3$. The development of desingu-larization theory has shown that this is more difficult than coarse desingularization of surfaces: it is more or less equivalent to smart desingularization of a surface in the special case where this surface is a hypersurface. Anyway simplification is known but we still face a serious problem. If $X \longrightarrow S$ is a ramified covering with $\dim S = 3$ and such that the discriminant Δ has normal crossings, we can argue as above and see that X is locally analytically a quotient singularity. The prototype of them are toroïdal embeddings. One can desingularize toroïdal embeddings of any dimension but, unfortunately, there are many ways of doing that and there is no known rule to choose one of them in such a way that the constructions made locally will patch together and give a global desingularization of X.

§ 4. De luxe JUNG's method.

ZARISKI has shown that JUNG's idea can be used to prove smart desingularization for a surface-hypersurface. We just saw that for a normal surface X, the existence of a finite projection $\pi : X \longrightarrow S$ with normal crossing discriminant allows us to give an explicit description of X as a union of toroïdal embeddings. ZARISKI'S method is based on the fact that for a surface which is a hypersurface the existence of such a projection implies that the problem of lowering the multiplicity is a purely combinatorial one. From that he is able to extract a proof of the smart desingularization theorem in that case: he says that you first have to blow up points, he says when you stop blowing up points and start blowing up ν- fold curves and of course he proves that you eventually get rid of all ν- fold points. Another version of an analogous idea has been given some years ago by ABHYANKAR, but I will not talk about it since it is going to be published in the proceedings of the Reinhardtsbrunn Conference 1978.*

Let us give the main features of ZARISKI's proof. The argument being limited to char.0, we avoid unessential difficulties by declaring that, up to the end of this lecture, scheme means of finite type over the complex numbers.

Definition Let X be a scheme.
We say that X is a hypersurface locally at x if rank $(\underline{M}_{X,x}/\underline{M}^2_{X,x}) = 1 + \dim O_{X,x}$. One has a surjective morphism of graded rings

$$(1) \qquad k[\underline{M}/\underline{M}^2] \longrightarrow \bigoplus_{n \geq 0} \underline{M}^n/\underline{M}^{n+1} \quad , \quad k = k(x) \ ,$$

we denote by $\operatorname{ord}_x(X)$ the smallest integer n such that (1) is not injective in degree n .

We know that if X is a hypersurface at a closed point x then there exists an etale neighbourhood U of X and a closed immersion $U \longrightarrow Z$ where Z is smooth over k and U is a Cartier divisor in Z .
From this we deduce that X is a hypersurface at all points x' belonging to the image of U and that $\operatorname{ord}_x(X)$ is semicontinuous. Hence, there exists a maximum ν for $\operatorname{ord}_x(X)$.

*Editor's Note: Abhyankar did not submit a paper to these proceedings. The reader may refer instead to the article by U.Orbanz in this volume, pp. 1-50.

A ν-fold scheme is a closed subscheme Y of X which is smooth and such that $ord_x(X) = \nu$ for any $x \in Y$; we will also say that Y is ν- permissible. One can prove that by blowing up a ν-fold Y in X we get an X' which is still a hypersurface and that $\nu(X') < \nu(X)$, see Prop.1 underneath. If we can prove that there exists a sequence of ν-permissible blowing up

$X = X_o \longleftarrow X_1 \longleftarrow X_2 \dots\dots X_{n-1} \longleftarrow X_n$ such that $\nu(X_n) < \nu(X)$ we obviously get smart desingularization by repeating this process a finite number of times (except for condition (iv)) .

Now the first point is that if one has a transversal projection $\pi: X \longrightarrow S$ and a ν-permissible center Y in X then after blowing up Y we still get a transversal projection. More precisely we have the following

Proposition 1. Let $\pi: X \longrightarrow S$ a flat and quasi-finite morphism. Assume that

(i) S is regular and $\dim S = 2$ and X is everywhere a hypersurface. Let $\nu = \nu(X) = \max\left\{ord_x(X) , x \in X\right\}$.

(ii) For any $x \in X$ such that $ord_x(X) = \nu$, one has $\operatorname{length}_{O_{X,x}} (O_{X,x}/ \underline{M}_{S,s}O_{X,x}) = \nu$, where $s = \pi(x)$.

Then if $x \in X$ and $\nu = ord_x(X)$ then locally for the etale topology there exists an open neighbourhood U of x in X such that

(A) If Y is a closed ν-permissible subscheme of U the projection π induces an isomorphism $\pi_o: Y \xrightarrow{\sim} \pi(Y) \subset S$; hence $\pi(Y)$ is smooth.

(B) By blowing up of U with center Y and of $\pi(U)$ with center $\pi(Y)$ one gets a commutative diagram $(\star)$ and if $\nu(U) = \nu(U')$ then (i) and (ii) hold for π' .

$$(*)\qquad \begin{array}{ccc} U & \longleftarrow & U' \\ {\scriptstyle\pi}\downarrow & & \downarrow{\scriptstyle\pi'} \\ \pi(U) & \underset{g}{\longleftarrow} & S' \end{array}$$

We leave to the reader the reduction to the following case:

$S = \mathrm{Spec}(A)$ and $X = \mathrm{Spec}(A[T]/fA[T])$, $f[T] = T^{\nu} + a_1T^{\nu-1} + .. + a_{\nu}$, $a_i \in A$.

We perform the usual trick $T \longmapsto T + \frac{a_1}{\nu}$, and we can assume that $a_1 = 0$.

Let $S_{\nu}(X) = \left\{ x \in X ,\ \mathrm{ord}_x(X) = \nu \right\}$

Then I claim that $T = 0$ at all points of $S_{\nu}(X)$ and this will obviously prove (A). Let us prove the claim. If x is a ν-fold point, by which I mean that $x \in S_{\nu}(X)$, then $f \in \underline{M}^{\nu}_{Z,x}$, where $Z = \mathrm{Spec}\,(A[T])$, hence $(\nu!).\,T = \frac{\partial^{\nu-1}}{\partial T^{\nu-1}} f \in \underline{M}_{Z,x}$ hence $T = 0$ at x. Now we are going to prove (B). Let Y be a ν-permissible center for X. By definition Y is a smooth subscheme of Z of codimension 2 (resp.3) hence locally around any point $x \in Z$ one can choose equations of Y in Z of the form (T,x_1) (resp. T,x_1,x_2). Such that the equations of $\pi(Y)$ in S are (x_1) (resp.(x_1,x_2)). In other words, if P is the ideal of Y in Z and Q the ideal of $\pi(Y)$ in S one has $P = QA[T] + TA[T]$, and $Q = P \cap A$. Since $f \in P^{\nu}$, then $\frac{\partial^k}{\partial T^k} f \in P^{\nu-k}$, $1 \leq k \leq \nu-1$, which implies $a_i \in Q^i$. We are now going to describe the diagram (*) of prop 1.

<u>First case dim Y = 1</u>. We know that X' is covered by two affine pieces. In the first one we have $PO_{X'} = TO_{X'}$ and in the second one $PO_{X'} = x_1O_{X'}$. Since $a_i \in x_1^iA$, the unitary equation $T^{\nu} + a_2T^{\nu-2} + .. + a_{\nu}$ can be written

$$T^{\nu} + x_1^2a'_2\,T^{\nu-2} + \cdots + x_1^{\nu}\,a'_{\nu}\ ,\ a'_i \in A\ ,$$

and from this follows that in fact X' is <u>equal</u> to the open set where $PO_{X'} = x_1O_{X'}$ hence X' is affine and we have

$$X' = \mathrm{Spec}\ A[T_1]/(T_1^{\nu} + a'_2\,T_1^{\nu-2} + \cdots + a'_{\nu})\ ,\ T_1 = T/x_1\ .$$

As a consequense X' is finite over X. On the other hand $S' = S$ since $\pi(Y)$ is a Cartier divisor, hence blowing up of $\pi(Y)$ amounts to nothing. Property (B) is obvious in that case.

<u>Second Case dim Y=0</u>, $Q = (x_1,x_2)$, $P = (T,x_1,x_2)$ and of course $Y = \{\xi\}$ is a closed point. The blown up X' is covered by three affine pieces, but as above one sees that it is covered already by only two of them, namely

$$X'(i) = \{\eta \in X' \;/\; x_i O_{X',\eta} = PO_{X',\eta}\},\; i = 1,2 .$$

The blown-up S' is also covered by two affine pieces

$$S'(i) = \{\eta \in S' \;/\; x_i O_{S',\eta} = Q\, O_{S',\eta}\}\; i = 1,2 ;$$

Let $S'(i) = \operatorname{Spec}(A(i))$. Since $a_n \in Q^n$ one has $a_n \in x_i^{\,n} A(i)$, hence there exist $a'_2, a'_3, \ldots, a'_\nu$ in $A(i)$ such that $a_2 = x_i^2 a'_2$, $a_3 = x_i^3 a'_3$, ..., $a_\nu = x_i^\nu a'_\nu$ and eventually

$$\text{for } i = 1,2, \quad \begin{cases} S'(i) = \operatorname{Spec} A(i) \\ X'(i) = \operatorname{Spec} A(i)[T_1]/(T_1^\nu + a'_2 T_1^{\nu-2} \ldots + a'_\nu) \\ T_1 = T/x_1 \end{cases}$$

Hence property (B) is proved and we even have an explicit description of the way the equation of X behaves by permissible blowing up. Furthermore, observe that the discriminant δ of $f(T)$ is a weighted homogeneous polynomial in the variables a_i with total weight $\nu(\nu-1)$. Hence if Q is the ideal of $\pi(Y)$ in S, since we know that $a_i \in Q^i$, we know that $\delta \in Q^{\nu(\nu-1)}$. After blowing up we know that $a_i \longmapsto a_i/t^i$ where t <u>is the equation of the exceptional divisor</u> in some affine piece of the blown up scheme S'. Hence we know that the new discriminant is

$$(\star) \qquad \delta' = \delta / t^{\nu(\nu-1)}$$

As a consequence, if δ had normal crossing in S then δ' also has normal crossing in S'. Proof: if we blow up one point this is clear;

if we blow up a ν-fold curve we do not change S .

<u>We start explaining how to use Prop. 1 .</u>

One can cover X by affine pieces X_α each of which is equipped with a projection $\pi_\alpha : X_\alpha \longrightarrow S_\alpha$ with the properties (i) and (ii) . For each of them we have a discriminant Δ_α , hence a finite set F_α in each X_α such that $x \in F_\alpha$ means that $\mathrm{ord}_x(X_\alpha) = \nu$ and $\pi(x)$ is a point of S_α where Δ_α is not normal crossings. If we blow up F_α in X we get a new family $X'_\alpha \longrightarrow S'_\alpha$ and the X'_α still cover X' . As we said in §2 , if we keep blowing up these points then the Δ_α will eventually become normal crossing divisors. And this property will be stable by any further ν-permissible blowing up . But now we will have to perform blowing up with one-dimensional support.

Here comes the fundamental lemma of Zariski which tells us what happens.

<u>Proposition 2.</u> Let S be smooth of dimension 2 and let $X = \mathrm{Spec}\, A[T]/(T^\nu + a_2T^{\nu-2} + \ldots + a_\nu)$, where $S = \mathrm{Spec}(A)$. Let u_1, u_2 in A such that $u_1u_2 = 0$ is a normal crossing divisor of S . Call this divisor Δ and let $S_0 = S - \Delta$. Assume that $X_0 = X \times_S S_0$ is etale above S_0 and that s is a point of S with $u_1 = u_2 = 0$.
Let Γ_1 be the curve with equation $T = u_1 = 0$
Let Γ_2 be the curve with equation $T = u_2 = 0$

Then $S_\nu(X) \subset \Gamma_1 \cup \Gamma_2$.

Furthermore there exist two numbers λ_1 and λ_2 such that if we replace S by a convenient neighbourhood of ξ we have

(i) $\lambda_1 > 1 \leftrightarrow \Gamma_1 \subset S_\nu(X)$

(ii) $\lambda_2 > 1 \leftrightarrow \Gamma_2 \subset S_\nu(X)$

(iii) $\lambda_1 + \lambda_2 > 1 \leftrightarrow \{\xi\} = \Gamma_1 \cap \Gamma_2 \subset S_\nu(X)$.

We already saw that $T = 0$ at any point of $S_\nu(X)$ because $a_1 = 0$. Furthermore $\pi(S_\nu(X))$ is contained in the discriminant locus Δ hence $S_\nu(X) \subset \Gamma_1 \cup \Gamma_2$.

Replace S by a small enough complex analytic neighbourhood of s ; there exists a complex analytic branched covering $S_N \longrightarrow S$, given by $u_1 = x_1^N$, $u_2 = x_2^N$ with <u>smooth</u> S_N and such that the covering X_0/S_0 is trivialized by the base change $S_N \longrightarrow S$ for instance let N be $\nu!$. In other words we have ν morphisms $r_\alpha : S_N \longrightarrow X$, each one of them being given by complex analytic functions

$$T = r_\alpha(X_1,X_2) \ , \ u_1 = x_1^N \ , \ u_2 = x_2^N \ ,$$

and we have $\Pi(T - r_\alpha(x_1,x_2)) = f(T,u_1,u_2)$.

Let $R = \text{G.C.D}\ (r_\alpha - r_\beta \ , \ \alpha \neq \beta)$.

Since the discriminant $\delta = \prod_{\alpha\neq\beta}(r_\alpha - r_\beta)$

is $\delta = \text{unit}.u_1^a u_2^b$, one has that $R = \text{unit}.\ x_1^{m_1} x_2^{m_2}$. We let

(1) $$\lambda_1 = m_1/N \ , \quad \lambda_2 = m_2/N \ .$$

I claim that there exists a complex analytic function $g(u_1,u_2)$ such that for each root $r_\alpha(x_1,x_2)$ one has

(2) $$r_\alpha(x_1,x_2) = g(u_1,u_2) + x_1^{m_1} x_2^{m_2} h_\alpha(x_1,x_2) \ .$$

Let $r_\alpha = \Sigma\, r_{\alpha,i,j}\, x_1^i\, x_2^j$ be a root and define

$$g_\alpha(u_1,u_2) = \sum_{N|i,\,N|j,\,(i<m_1 \text{ or } j<m_2)} r_{\alpha,i,j}\, x_1^i\, x_2^j \ .$$

Since $x_1^{m_1}\, x_2^{m_2}$ divides $r_\alpha - r_\beta$ we get that g_α does not depend on α. Call it $g(u_1,u_2)$.

To prove (2) we introduce the Galois group G of the covering S_N/S : this group is generated by

$(x_1,x_2) \longmapsto (\xi x_1,x_2)$ and $(x_1,x_2) \longmapsto (x_1,\xi x_2)$

where ξ is a primitive N-root of unity. Since G acts on the roots, if we look at a monomial $r_{\alpha,i,j}\, x_1^i\, x_2^j$ with $N \nmid i$ and $r_{\alpha,i,j} \neq 0$, then we must have $i > m_1$ and $j > m_2$, since $r_\alpha(\xi x_1, x_2)$ is a root ; idem if $N \nmid j$ and this proves (2) .

We now prove the $\Rightarrow$ part of (i) . If $\lambda_1 > 1$, then $m_1 > N$ and since $f(T) = \prod_{1 \leq \alpha \leq \nu} (T-g-x_1^{m_1} x_2^{m_2} h_\alpha)$ we have that $f(T)$ belongs to the ideal $(T-g, x_1^N)^\nu$ hence the curve Γ_1' with equation $T-g = u_1 = 0$ is ν-permissible. As we saw that $S_\nu(X) \subset \Gamma_1 \cup \Gamma_2$ we have $\Gamma_1' = \Gamma_1$ hence Γ_1 is ν-permissible and u_1 must divide g .

Similarly for the $\Rightarrow$ part of (iii), we have $\lambda_1 + \lambda_2 > 1$ hence $m_1 + m_2 > N$ hence $f \in (T-g, (x_1,x_2)^N)^\nu$ hence $f \in (T-g, u_1, u_2)^\nu$. This exactly means that the point ξ' defined by $T-g = u_1 = u_2 = 0$ is ν-fold. As we already saw that $T = 0$ at any ν-fold point we conclude that $\xi' = \xi$ where ξ is defined by $T = u_1 = u_2 = 0$, in other words $\{\xi\} = \Gamma_1 \cap \Gamma_2$, hence $\Rightarrow$ of (iii) is true.

The proof of the converse is very tricky and since we don't find a better one we refer to [Zariski O ., Exceptional singularities of an algebroid surface and their reduction. Acad. Naz. dei lincei, Serie VIII, Vol XLIII, fasc.3-4-(1967) p.135-146] which is reproduced in volume I of [ZARISKI's collected papers, M.I.T. Press (1972)].

The same paper also describes the behavior of λ_1 and λ_2 under blowing up with ν-fold center which is as follows

(A) if $\lambda_1 > 1$ and if we blow up Γ_1 then we don't change S and don't change Γ_1 and Γ_2 but (λ_1, λ_2) is replaced by $(\lambda_1 - 1, \lambda_2)$

(B) if $\lambda_1 + \lambda_2 > 1$ and if we blow up $\{\xi\} = \Gamma_1 \cap \Gamma_2$ then the d.n.c. will acquire a new component namely the exceptional divisor of the blowing up of $s = \pi(\xi)$ in S. Call it Γ_e. The three numbers $(\lambda_1, \lambda_e, \lambda_2)$ attached to $\Gamma_1, \Gamma_e, \Gamma_2$ will be $\lambda_1, \lambda_1 + \lambda_2 - 1, \lambda_2)$. Picture:

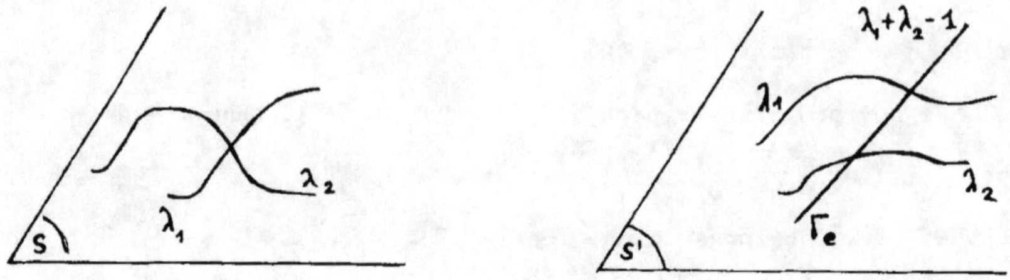

We will not say how this local study is used by ZARISKI to get the desingularisation of surfaces for two reasons. The first one is that we will mimic this part of his argument in the next § . The second one is that one can find ZARISKI's proof in [ZARISKI O., A new proof of the total embedded resolution theorem for algebraic surfaces (based on the theory of quasi-ordinary singularities), Amer.J. of Math. (1978)]. Since this paper is fairly cleverly organized it is not to be explained in a few pages, and the interested reader should look at it. He will see that these two papers deserve the qualification of "de luxe".

§ 5. HIRONAKA's method applied to surfaces.

As ZARISKI does in the paper we have just quoted, we will reduce the problem to a d.n.c. case, but here this condition does not refer to any kind of discriminant but rather to the coefficients of the equation of the singularity. Hence we will not use Prop. 2 of § 4. Here surface means reduced scheme purely of dimension 2 and of finite type over $\mathbb{C}$. We will only treat the hypersurface case and refer to remark 2 at the end of the lecture for the general case.

We will repeatedly use in the proofs the local model we already studied in § 4 , namely:

S is a smooth surface,

$Z = \operatorname{Spec} O_S[T]$

$X = \operatorname{Spec} (O_S[T]/f(T))$

With $f(T) = T^\nu + a_2T^{\nu-2}+..+a_\nu$, $a_i \in O_S(S)$.

As above if P is the ideal in Z of a ν-fold subscheme Y of X one must have $T \in P$ and $a_i \in Q^i$, where $Q = P \cap O_S$.

Proof: $\frac{\partial^k f}{\partial T^k} \in P^{\nu-k}$. Conversely if Q is an ideal of O_S such that

(2) $\operatorname{Spec} (O_S/Q)$ is smooth and $a_i \in Q^i$, $i = 2,3,\ldots,\nu$,

then $P = Q\, O_Z + T \cdot O_Z$ is the ideal in Z of a ν-fold subscheme Y of X .

Hence the whole process of ν-permissible blowing up can be expressed in terms of S and the a_i . After blowing up, the a_i's are only locally defined modulo the choice of a generator of the ideal of the exceptional divisor. To avoid this difficulty we introduce ideals

(3) $A_i = a_i^{\nu!/i} O_S$, $i = 2,3,..,\nu$, and we know that if Q is the ideal of a smooth closed subscheme Y' of S then the subscheme Y of Z whose ideal is $T\, O_Z + Q\, O_Z$ will be a ν-fold subscheme of X if and only if

(4) $Q^{\nu!} \supset A_i$, $i = 2,3,..,\nu$.

Furthermore if S' is the blowing up of S with center Y' (which is equal to S if Y' is a curve) then

(5) A_i is replaced by $Q^{-\nu!} A_i O_{S'} = A'_i$.

(A) <u>Normal crossing condition</u>

We say that we have <u>normal crossing</u> at $s \in S$ if all the ideals $A_2, A_3, \ldots, A_\nu$, $\prod_{2 \le i \le \nu} A_i$, $\sum_{2 \le i \le \nu} A_i$ are d.n.c.

in a neighbourhood of s . In that case there exists an integer $i \in [2,\nu]$ such that for any $j \in [2,\nu]$, $A_i \supset A_j$ (in other words A_i divides A_j)

in a neighbourhood U of s . Hence an ideal Q is ν-fold in U if and only if $Spec(O_U/Q\,O_U)$ is smooth and $A_i \subset Q^{\nu!}$ (in U) . Furthermore the corresponding $Y' \cap U$ has to be either a branch of the d.n.c or a point belonging to the d.n.c. and after blowing it up we will have again normal crossing. Furthermore for any point $s' \in S'$ lying over $s \in S$ the same index i will be such that $A'_i \supset A'_j$ for any $j \in [2,\nu]$ (See(5)) .

If s is a point of the d.n.c. we will have local coordinates (x_1,x_2) and integers (λ_1, λ_2) such that $A_i O_{S,s} = (x_1^{\lambda_1}\, x_2^{\lambda_2})\, O_{S,s}$.

If the center of blowing up Y' is a curve, its equation will be for instance $x_1 = o$ (locally at S) we will have S' = S and

$$A'_i\, O_{S,s} = (x_1^{\lambda_1 - \nu!},\ x_2^{\lambda_2})\, O_{S,s}$$

If the center of blowing up Y' is a branch point, let $\pi : S' \longrightarrow S$ be the blowing up and let $U' = \pi^{-1}(U)$. In U' we will only have two branch points lying on the exceptionnal divisor with the following rule ;

(i) let s_1 be the branch point lying above s and on the strict transform of the curve $x_1 = o$. We have local coordinates $t = x_2$ and $x'_1 = x_1/t$ with

$$A'_i\, O_{S',s_1} = (x'^{\lambda_1}_1,\ t^{\lambda_1 + \lambda_2 - \nu!})\, O_{S',s_1}$$

(ii) let s_2 be the branch point lying on the strict transform of the curve $x_2 = o$. We have local coordinates $t = x_1$, $x'_2 = x_2/t$ with

$$A'_i\, O_{S',s_2} = (t^{\lambda_1 + \lambda_2 - \nu!},\ x'^{\lambda_2}_2)\, O_{S',s_2} .$$

We will use later on the following

Observation. ($\lambda_1 < \nu!$ and $\lambda_2 < \nu!$) $\leftrightarrow$ (there is no ν-fold curve through s) . In that case let $\mu = \lambda_1 + \lambda_2$. Then ($\mu > \nu!$) $\leftrightarrow$ (s is ν-fold). If we blow up s when $\lambda_1 < \nu!$, $\lambda_2 < \nu!$, $\mu > \nu!$, then the exceptionnal divisor is not ν-fold since the corresponding λ is $\lambda_1 + \lambda_2 - \nu!$. Furthermore, for the two branch points s_1 and s_2 lying above s we have the numerical caracters

$$(\lambda_1, \lambda_2, \mu) = \begin{cases} (\lambda_1, \mu-\nu!, \mu + \lambda_1 - \nu!) & \text{at } s_1 \\ (\lambda_2, \mu-\nu!, \mu + \lambda_2 - \nu!) & \text{at } s_2 \end{cases} .$$

Since λ_1, λ_2 are $< \nu!$ one has $\mu < 2\,\nu!$ hence the hypotheses ($\lambda_1 < \nu!$ and $\lambda_2 < \nu!$) is still true at s_1 and s_2 . Furthermore one has $\mu(S_1) < \mu(s)$ and $\mu(s_2) < \mu(s)$.

Remark. If one knew that « normal crossing condition » does not depend of the local model we choosed at $\xi \in X$, we could blow up repeatedly the discrete, finite, closed set where this condition is not satisfied. As we said earlier since dim $S = 2$ we would reach normal crossing after a finite number of steps. Then it would be a purely combinatorial game to get rid of all ν-fold points or curves. This is the line of reasoning of Hironaka when he does his gardening. For the sake of simplicity, we are going to argue differently but we will describe an algorithm with the following properties

(i) if $X' \longrightarrow X$ is an etale morphism, if $P(X)$ is the result of the algorithm applied to X then $X' \times_X P(X)$ is $P(X')$.

(ii) one can describe the algorithm without having to know anything about the way we prove that it works.

(B) Blowing up of ν-fold curves

Lemma 1. Let X be a surface-hypersurface and let C be an irreducible ν-fold curve with $\nu = \nu(x)$. Let $p: X' \longrightarrow X$ be the blowing up of X with center C. Then p is a finite morphism. Furthermore, if

(*) $\quad \dim (S_\nu(X') \cap p^{-1}(C)) > 0$

then $p^{-1}(C)_{red}$ is smooth and the projection π induces an isomorphism $p^{-1}(C)_{red} \xrightarrow{\sim} C$.

Let ξ be a closed point of C. There exists a complex analytic neighbourhood U of ξ in X provided with a finite and flat projection

(1) $\quad \pi: U \longrightarrow S$

such that S is smooth, and U is defined in $Z = S \times \mathbf{A}^1$ by an equation

(2) $\quad f(T) = T^\nu + a_2 T^{\nu-2} + \ldots + a_\nu$

the a_i being holomorphic functions on S. Furthermore one knows that $T = 0$ on C hence π induces an isomorphism $C \xrightarrow{\sim} \pi(C)$. Let $t = 0$ be the equation of $\pi(C)$ (choose a smaller U if necessary). According to proposition 1 of § 4 we have a diagram

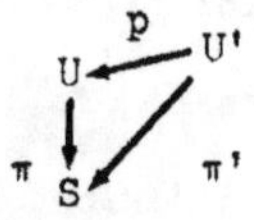

where U' is the blowing up of U with center $U \cap C$. As we saw in the proof of the proposition 1 of § 4, we have that p is a finite morphism. Let $T_1 = T/t$ be the variable in U'; we know that $T_1 = 0$ on $S_\nu(U')$ hence $\pi'(S_\nu(U')) \simeq S_\nu(U')$. It follows that the projection π' induces an embedding of $S_\nu(U') \cap p^{-1}(C)$ into $\pi'(p^{-1}(C)) = \pi(C)$.
Hence if

(*) $\quad \dim (S_\nu(U') \cap p^{-1}(C)) > 0$

one has that $\pi'(S_\nu(U') \cap p^{-1}(C)) = \pi(C)$.

[We should have chosen U such that $C \cap U$ is connected and warned that C has been replaced by $C \cap U$].

The lemma follows from this. One could avoid the use of complex analytic geometry by considering the completion of $O_{X,\xi}$ or an etale neighbourhood of ξ ; it is a matter of taste.

(C) Blowing up of a closed point.

Lemma 2 Let ξ be a ν-fold point of X with $\nu = \nu(X)$. Let $p: X' \longrightarrow X$ be the blowing up of ξ .

Assume that

(*) $\qquad \dim (S_\nu(X') \cap p^{-1}(\xi)) > 0$

then $p^{-1}(\xi)_{red}$ is a projective line contained in $S_\nu(X')$.

Again we choose a local analytic model $\pi: X \longrightarrow S$ and blow up ξ in X and $s = \pi(\xi)$ in S . We get a commutative diagram as in §4 .

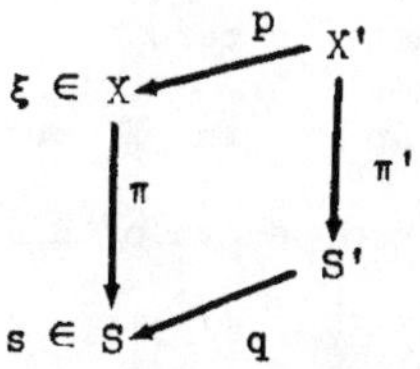

We know that π' maps $S_\nu(X')$ isomorphically on its image. Hence $\pi'(S_\nu(X') \cap p^{-1}(\xi))$ is mapped isomorphically on a closed subscheme of $q^{-1}(s)$. Since $q^{-1}(s)$ is a projective line this image is either a finite set of closed points or $q^{-1}(s)$ itself Since we have (*) we are in the second case. Hence we get the conclusion. Picture:

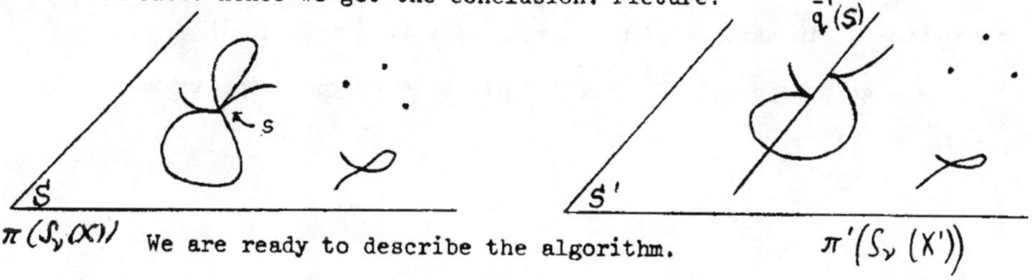

We are ready to describe the algorithm.

<u>Operation 1</u>. <u>Datum</u>: X where X is a reduced scheme of finite type over $\mathbb{C}$. X is purely of dimension 2 and for any $x \in X$ one has rank $(\underline{m}_{X,n} / \underline{m}_{X,n}^2) = 1 + \dim O_{X,n}$. For short we say that X is a ≺ surface-hypersurface≻ .

Let $\nu = \max \left\{ \operatorname{ord}_x(X) \ , \ x \in X \right\}$. IF $\nu = 1$ GO TO END .

Let $S_\nu(X) = \left\{ x \in X \mid \operatorname{ord}_x(X) = \nu \right\}$ with the reduced structure

$$I_\nu(X) = \left\{ x \in X \mid x \text{ is isolated in } S_\nu(X) \right\}$$

$$\Sigma_\nu(X) = S_\nu(X) \setminus I_\nu(X) \ (= 1\text{-dimensional part of } S_\nu(X))$$

$$F_\nu(X) = \operatorname{Sing}(\Sigma_\nu(X))$$

IF $F_\nu(X) \neq \phi$ apply Operation 2 to $(X, F_\nu(X), \nu)$.

IF $F_\nu(X) = \phi$. IF $\Sigma_\nu(X) \neq \phi$ apply operation 3 to $(X, \Sigma_\nu(X), \nu)$.

IF $F_\nu(X) = \phi$, IF $\Sigma_\nu(X) = \phi$ apply operation 2 to $(X, I_\nu(X), \nu)$.

<u>Operation 2 .</u> Data: (X,F,ν) where X is a surface -hypersurface, F is a reduced closed subscheme of dimension zero of X , any point of F is ν-fold and $\nu = \nu(X)$. Blow up F in X and get $p: X' \longrightarrow X$. Look at the union of those irreducible components of dimension 1 of $p^{-1}(F)_{red}$ wich are ν-fold, call this union C' . IF $C' = \phi$ go to operation 1 for X' . If $C' \neq \phi$ go to operation 3 for (X', C', ν) .

<u>Operation 3 .</u> <u>Data</u>: (X,C,ν) where X is a surface-hypersurface, C is a ν-fold curve(not necessarily connected) and $\nu = \nu(X)$.
Blow up C and get $p: X' \longrightarrow X$. Let C' be the union of the irreducible components of dimension 1 of $p^{-1}(C)red$ wich are ν-fold . IF
IF $C' = \phi$ go to operation 1 for X' . IF $C' \neq \phi$ go to operation 3 for (X', C', ν)

We first have to check that when we go from one operation to another one, the new data satisfy the hypothesis of the second operation. Since permissible blowing up preserve the surface-hypersurface condition the only difficulty may occur when we go from 2 to 3 or 3. In both cases one still has $\nu = \nu(X)$ since $C' \neq \phi$ (otherwise we would go to 1). By definition each irreducible component of C' is smooth since it is ν-fold. On the other hand, according to (B) and (C) there is at most <u>one</u> irreducible component of C' above each irreducible component of the center of blowing up Y (which is F for operation 2 or C for operation 3). Since Y is smooth, two distinct irreducible components of Y do not meet hence two distinct irreducible components of C' do not meet hence C' is smooth and obviously ν-fold.

<u>Property 1</u>. Let $\pi: X' \longrightarrow X$, be obtained from (X,Y,ν) by performing operation 2 (resp. 3), with $Y = F$ (resp. $Y = C$). It is clear that $\Sigma_\nu(X')$ contains the strict transform $\Sigma_\nu(X)_{st}$ of $\Sigma_\nu(X)$. Let E be the union of those irreducible components of $\Sigma_\nu(X')$ which are not contained in $\Sigma_\nu(X)_{st}$. We have that $\pi(E) \subset S_\nu(X)$. In the case of operation 2 we must have that $\pi(E) \subset F$ hence according to (B), E is the union of a finite set of mutually disjoint projective lines. Hence $E = C'$. In the case of operation 3, for each irreducible component E_α of E one has $\dim \pi(E_\alpha) = 1$ since $\pi: X' \longrightarrow X$ is a finite morphism. Since $E_\alpha \not\subset \Sigma_\nu(X)_{st}$ one must have that $E_\alpha \cap C$ is dense in E_α (since $\pi(E_\alpha) \subset \Sigma_\nu(X)$ and $\pi(E_\alpha)_{st} = \phi$). Hence $\pi(E_\alpha)$ is a component of C. Hence E_α is ν-fold as we saw in (B), hence $\bigcup_\alpha E_\alpha \subset C'$ and obviously $\bigcup_\alpha E_\alpha = C'$. As a conclusion we get

(D) $$\Sigma_\nu(X') = \Sigma_\nu(X)_{st} \cup C'$$

with $\Sigma_\nu(X)_{st} \cap C'$ is finite and C' is a union of mutually disjoint ν-fold irreducible curves.

<u>Property 2.</u> One can only perform operation 3 a finite number of times before having to go back to operation 1.

Remember that repeating operation 3 produces a succession of finite and birationnal morphisms $X = X_0 \longleftarrow X_1 \longleftarrow X_2 \dots X_n$. By the finiteness of the normalization of X one only has to show that $X_i \neq X_{i+1}$ for $i = 0,1,..,n-1$. In other words, with the notation of operation 3, the morphism $\pi : X' \longrightarrow X$ cannot be an isomorphism. Since ν can only change during operation 1, one has $\nu \geq 2$ hence X is not smooth at any point $\xi \in C$; since C is smooth it follows that C is not a Cartier divisor of X hence π is not an isomorphism (check that $C \neq \phi$). Alternate proof: look at what happens in the basis S of any local model.

<u>Property 3.</u> When performing operation 1 for some X, when we first go back to operation 1 we find some X' and we have that $\Sigma_\nu(X') = \Sigma_\nu(X)_{st}$ is the strict transform of $\Sigma_\nu(X)$.

By transitivity of the strict transform this follows from (D). In fact when performing 2 or 3, if $C' \neq \phi$ we go to operation 3 and we can only perform operation 3 a finite number of times, say n times, before having to go back to operation 1. This means that we have a sequence of modifications

$$X \longleftarrow X_1 \longleftarrow X_2 \dots X_{n-1} \longleftarrow X_n$$

and ν-fold curves C_1 in X_1, C_2 in $X_2, \dots, C_n$ in X_n such that for $i = 1,2,-,n-1$, X_{i+1} is obtained by blowing up X_i with center C_i. Furthermore $C_n = \phi$ expresses the fact that we have to go back to operation 1. Formula (D) tells us that $\Sigma(X_1) = \Sigma(X)_{st} \cup C_1$ hence $\Sigma(X_2) = \Sigma(X)_{st} \cup C_{1,st} \cup C_2$ but the strict transform of the center of blowing up C_1 is empty hence $\Sigma(X_2) = \Sigma(X)_{st} \cup C_2$ and by induction $\Sigma(X_n) = \Sigma(X)_{st} \cup C_n = \Sigma(X)_{st}$.

Property 4. After a finite number of passages through operation 1 one has $\Sigma_\nu(X) = \phi$ which means $S_\nu(X) = I_\nu(X)$. Furthermore once this is achieved this remains true as long as $\nu(X)$ remains constant.

Remark that if $F_\nu(X) \neq \phi$ then, with notation of property 3 $\Sigma_\nu(X') = \Sigma_\nu(X)_{st}$ and at any singular point of $\Sigma_\nu(X)$ we have performed some non-trivial blowing up. Hence after a finite number of passages $\Sigma_\nu(X)_{st}$ is smooth hence ν-fold. Then operation 1 tells us to go to 3 which means that the new strict transform will be empty and we get the conclusion by (D).

To prove that the algorithm stops we can assume that $I_\nu(X) = S_\nu(X)$ and we only have to show that ν eventually drops. Since $I_\nu(X)$ is finite we only have to look at some $\xi \in I_\nu(X)$ and we can replace X by some local model $\pi : X \longrightarrow S$, $\pi(\xi) = s$, as in (A) . We call

(1) $$X = X_0 \longleftarrow X_1 \longleftarrow X_2 \ldots X_{n-1} \longleftarrow X_n \ldots$$

the result of the successive passages through operation 1, and we have to show that for n big enough one has $\xi \notin p(S_\nu(X_n))$.

Let $\xi_n \in S_\nu(X_n) = I_\nu(X_n)$

and call $\xi_{n-1}, \ldots \xi_1, \xi_0 = \xi$ it's successive projections. Since operation (3) does not change S (blowing up of a smooth curve) , we get a succession of blowing up with finite and reduced centers G_i in S called

$$S = S_0 \longleftarrow S_1 \longleftarrow S_2 \ldots S_{n-1} \longleftarrow S_n .$$

Let $s_n, s_{n-1}, \ldots, s_1, s_0 = s$ be the successive projections of the ξ_i .

Since we always pass through operation 2 between X_i and X_{i+1} , we know that $s_i \in G_i$. As a consequence of simplification of boundary in S (here $\dim S = 2$) we know that there exists N such that for $n > N$, we will have the d.n.c. condition at any point of G_n .

With the notations of (A) at the point s_n we know that $\lambda_1 < \nu !$ and $\lambda_2 < \nu !$ because we know that $S_\nu(X_n) = I_\nu(X_n)$ is of dimension 0 .

Hence $\mu(s_n) = \lambda_1 + \lambda_2 < 2\nu!$. But we also know that $\mu(s_{n+1}) < \mu(s_n)$ hence for $n > N + \nu!$ we will certainly have $S_\nu(X_n) = \phi$ since $\mu(s_n) < \nu!$.

This proves that the algorithm stops.

Remark 1. We have proven the smart desingularization theorem except for the condition (iv) which relates to normal crossing of the exceptional divisor. It is an easy exercise to modify operation 1 in such a way that the algorithm also achieves that.

Remark 2. If we only assume that X is reduced and purely of dimension 2 we let $\nu_x(X)$ be the HILBERT-SAMUEL series of X at x ,

$\nu_x(X) = \sum_{n \geq 0} \lg(M^n_{X,x}/M^{n+1}_{X,x})\, T^n$. We modify operation (1) as follows:

let ν be the maximum of $\nu_x(X)$. If $\nu = (1-T)^{-2}$ GO TO END.

let $S_\nu(X) = \{x \in X \mid \nu_x(X) = \nu\}$ with the reduced structure.

We don't change the sequel.

In the text of operation 2 and 3 replace «surface hypersurface» by X is reduced and purely of dimension 2.

I claim that this algorithm also stops. Of course the proof is more complicated due to the fact that the local models must be obtained by the use of a normalized standard basis for the ideal of a (local) embedding $X \subset Z$. Furthermore one has to show that ν cannot decrease indefinitely. This two difficulties are not easier to overcome for $\dim(X) = 2$ than for the general X . But when this is done, the proof of (A),(B),(C) and of the finiteness of the algorithm (for dim X = 2) are not seriously affected by the lack of the hypersurface hypothesis.

Remark 3. The algorithm can be changed in a few different ways and still enjoy the property (i) we stated at the beginning of this § . This is a good exercice for the interested reader.

DESINGULARIZATION IN DIMENSION 2

VINCENT COSSART
Université Pierre et Marie Curie (Paris VI)
Mathématiques
4, Place Jussieu
F - 75005 PARIS

INTRODUCTION. The aim of this paper is to link four different methods of desingularization of excellent surfaces: The methods of Jung, Zariski, Abhyankar and Hironaka. The basic reference is Giraud's lecture in this volume ([8]). In this paper, Giraud shows how close the methods of Zariski and Jung are. So we are going to show that the proofs by Zariski, Abhyankar and Hironaka are almost the same from the point of view of the characteristic polyhedron of singularity. Indeed these three authors want to reach this case: At every closed point of the worst Samuel stratum of X (= a reduced hypersurface of an excellent, regular scheme Z of dimension 3) there exists a regular system of parameters (y,u_1,u_2) such that the polyhedron $\Delta(J,u_1,u_2)$ (see [9] , 1.12)) has only one vertex, J being the ideal of X in $O_{Z,x}$. In that case, Hironaka calls the singularity "quasi-ordinary" and Abhyankar calls it "curve-like".

The first section defines properly the notion "$\Delta(J,u_1,u_2)$ has only one vertex" and shows how nice it is.

In the second section we translate Zariski's method into terms of the characteristic polyhedron.

In the third section we give a new proof of Abhyankar's theorem with the help of the characteristic polyhedron.

The fourth section is just a reminder of Hironaka's proof.

Acknowledgement

I would like to thank U. Orbanz for his thorough and critical reading of the drafts of these notes. He also carried out considerable editing on them and ultimately wrote them up in a much better final form than my original version.

I Quasi-ordinary points

To simplify the proofs of this paper, we will only look at the surface-hypersurface case. Indeed the proofs are long enough and the specialists will be convinced that most of the computations may be generalized to the case of a surface X embedded in a regular excellent scheme Z of any dimension.

NOTATION AND DEFINITION. Let us denote by X a reduced hypersurface of an excellent, regular scheme Z of dimension 3. We denote by $\mathrm{Sam}(X)$ the locus of highest multiplicity of X. Let $x \in \mathrm{Sam}(X)$ be a closed point and let $I(X) = (f)$ at x. We say that x is a quasi-ordinary point if it satisfies the two conditions:

(1) $\dim D_x(X) = 2$, where $D_x(X)$ is the directrix of the tangent cone $C_{X,x}$ of X at x. ([7], 1-31; this condition has the following meaning: If $\nu = \mathrm{ord}_x(f)$, then $\mathrm{in}_x(f) = \varepsilon . Y^{\nu}$ for some Y and some unit ε.)

(2) There exists a regular system of parameters (y,u_1,u_2) of Z at x such that: $(\mathrm{in}_x(y)) = I(D_x(X)) \subset \mathrm{gr}_x(O_Z)$, the polyhedron $\Delta(f;u_1,u_2;y)$ is minimal and has only one vertex (see [9], (3.1)).

REMARK. In the case where $O_{Z,x}$ and its residual field k have the same characteristic, the choice of u_1,u_2 defines a projection $\hat{X} \to \mathrm{Spec}(k[[u_1,u_2]])$, where $\hat{X} = \mathrm{Spec}\, \hat{O}_{X,x}$ is the completion of X at x. Therefore, we recognize Jung's idea: We project X onto a regular surface.

PROPOSITION 1. Assume that x is a quasi-ordinary point of X. Let (y,u_1,u_2) be parameters satisfying (2), let (λ_1,λ_2) be the vertex of $\Delta(f;u_1,u_2)$ and let $\nu = \mathrm{ord}_x(f)$. Then we have:

a) If $D_1 = V(y,u_1)$ is permissible for X (see [12], p. 107) if we blow up D_1, then there is at most one point x' near to x. Let Z' be the transform of Z. Then $(y' = y/u_1,u_1,u_2)$ is a regular system of parameters of Z' at x'. Furthermore, if x' is very near to x*), then it is a quasi-ordinary point and $\Delta(f/u_1^{\nu};u_1,u_2)$ has only one vertex with coordinates $(\lambda_1 - 1,\lambda_2)$.

b) If neither $V(y,u_1)$ nor $V(y,u_2)$ is permissible for X and if we blow up x, then there are at most two points near to x. Let

---

* Recall the notion of near and very near points to X : A point x' in some blowing up x is called near to x if at x' the multiplicity is the same as at x. x' is called very near to x, if in addition the dimension of the directrix is unchanged ([6], introduction).

Z' be the transform of Z. Then $(y' = y/u_1, u_1, u_2' = u_2/u_1)$ or $(y'' = y/u_2, u_1'' = u_1/u_2, u_2)$ is a regular system of parameters of Z' at the near points of x. If one of them is very near to x, say with parameters (y', u_1, u_2'), then it is quasi-ordinary and $(f/u_1^{\nu}, u_1, u_2')$ has only one vertex with coordinates $(\lambda_1 + \lambda_2 - 1, \lambda_2)$.

Proof. a) If D_1 is permissible and if we blow it up, then it is very well known that there is at most one near point x' to x and that $(y' = y/u_1, u_1, u_2)$ is a regular system of parameters of Z' at x' ([12], p.107). We also know that $\Delta(f/u_1^{\nu}; u_1, u_2; y')$ has only one vertex v with coordinates $(\lambda_1' - 1, \lambda_2)$. Let v denote the vertex of $\Delta(f; u_1, u_2)$ and let

$$\mathrm{in}_v(f) = Y^{\nu} + \sum_{i=1}^{\nu} \alpha(i) Y^{\nu-1} U_1^{i\lambda_1} U_2^{i\lambda_2}, \quad \alpha(i) \in k$$

where $Y = \mathrm{in}_x(y)$, $U_j = \mathrm{in}_x(u_j)$ and k is the residual field at x. (For the definition of $\mathrm{in}_v(f)$ see [9], (3.7)). Then

$$\mathrm{in}_{v'}(f/u_1^{\nu}) = Y'^{\nu} + \sum_{i=1}^{\nu} \alpha(i)\, Y'^{\nu-i} U_1'^{(\lambda_1 - 1)} U_2'^{i\lambda_2}$$

where $Y' = \mathrm{in}_{x'}(y')$, $U_j' = \mathrm{in}_{x'}(u_j)$. As $\mathrm{in}_v(f)$ is not a νth power, also $\mathrm{in}_{x'}(f/u_1^{\nu})$ is not a νth power, so v' is not a solvable vertex of $\Delta(f/u_1^{\nu}; u_1, u_2; y')$ (see [9], (3.8) and (3.9)).

b) If neither $V(y, u_1)$ nor $V(y, u_2)$ is permissible, then $\lambda_1 < 1$, $\lambda_2 < 1$ and $\lambda_1 + \lambda_2 > 1$. If we blow up x, it is very well known that the near points are in the two open sets O_1 (resp. O_2) where u_1 (resp. u_2) is the equation of the exceptional divisor. Let us write $F = \mathrm{in}_{\delta}(f)$ (see [5], (6) or [6], Remarks on p. 27, for definition). Then, with obvious notation,

$$F = Y^{\nu} + \sum_{i=1}^{\nu} \alpha(i)\, Y^{\nu-1} U_1^{i\lambda_1} U_2^{i\lambda_2}, \quad \alpha(i) \in k$$

Suppose that x' is a near point to x in O_1. Then at x' the strict transform X' of X is defined by

$$f/u_1^{\nu} = y'^{\nu} + \sum_{i=1}^{\nu} y'^{\nu-i} u_1^{i(\lambda_1+\lambda_2-1)} [u_2'^{i\lambda_2} \cdot A(i) + u_1 B(i)]$$

where $u_2' = u_2/u_1$ and where $A(i) \in O_{Z',x'}$ has residue $\alpha(i)$ in the residue field at x'. If now u_2' is invertible at x', then $\mathrm{ord}_{x'}(f/u_1^{\nu}) < \nu$ since $0 \leq \lambda_1 + \lambda_2 - 1 < 1$, in contradiction to the assumption that x' is near to x. Therefore (y', u_1, u_2') is a regular system of parameters at x' and $\Delta(f/u_1^{\nu};u_1,u_2';y')$ has only one vertex with coordinates $(\lambda_1+\lambda_2-1,\lambda_2)$. Finally this polyhedron is minimal because not both coordinates of the vertex can be integers ([9], (3.9.2)).

PROPOSITION 2. Assume that x is a quasi-ordinary point of X and let (y,u_1,u_2) be a regular system of parameters satisfying (2). Then the following algorithm stops: We blow up any one of the smallest permissible ideals (y,u_1), (y,u_2) or (y,u_1,u_2) and we do the same in the successive transforms for any very near point of x and its regular system of parameters constructed in Proposition 1.

Proof. Let (λ_1,λ_2) be the coordinates of the vertex of $\Delta(f;u_1,u_2)$. Then by [5], (4-2) or [6], Remarks on p. 27, we have $\delta(X,x)=\lambda_1+\lambda_2$. For any very near point x' of x, Proposition 1 shows that $\delta(X',x')<\delta(X,x)$. Therefore the result follows from the fact that $\delta(X,x)\in(1/\nu!)\cdot\mathbb{N}$ ([9], (3.6)).

PROPOSITION 3. Assume that x is a quasi-ordinary point of X and let (y,u_1,u_2) be a regular system of parameters satisfying (2). If D is any equimultiple curve through x, then $D=V(y,u_1)$ or $D=V(y,u_2)$. In particular D is regular at x.

Proof. Let X' be the transform after blowing up x and let D' denote the strict transform of D. If $D'=\emptyset$, then it is easy to see that $D=V(y,u_1)$ or $D=V(y,u_2)$. If $D'\neq\emptyset$, then there is a unique point x' on D' near to x, and by Proposition 2 we may assume that x' is not very near to x. Since D' is an equimultiple curve through x', we conclude that $\dim D_{x'}(X')=1$ ([12], p.117). Furthermore, if (y',u_1',u_2') is the regular system of parameters of Z' at x' given by Proposition 1, we get that $\Delta(f';u_1',u_2';y')$ has only one vertex, with coordinates (λ_1',λ_2') say, where f' is the equation of X' in Z'. Now $\dim D_{x'}(X')=1$ implies $\lambda_1'=0$ or $\lambda_2'=0$ ([12], p.117) so we may assume $\lambda_1'=0$, $\lambda_2'=1$. Now after blowing up $V(y',u_2')$ there is no more point near to x', which implies $D'=V(y',u_2')$. Projecting onto X we obtain $D=V(y,u_2)$. Similarly, if $\lambda_1'=1$ and $\lambda_2'=0$ then $D=V(y,u_1)$.

PROPOSITION 4. Let $\nu(X)$ denote the highest multiplicity of points on X, and assume that X satisfies:

(*) Every closed point $x\in \mathrm{Sam}(X)$ with $\dim D_x(X)=2$ is quasi-ordinary.

Consider any sequence of surfaces $X(n)$ such that $X(0) = X$ and $X(n)$ is a blowing up of $X(n-1)$ centered at any closed subset of $Sam(X(n-1)$. Then for n large enough we have

$$\nu(X(n)) < \nu(X) .$$

Proof. Suppose we have made n blowing ups

$$X = X(0) \leftarrow X(1) \leftarrow X(2) \leftarrow \dots \leftarrow X(n)$$

and $\nu(X(n)) = \nu(X)$. If x is an isolated point of $Sam(X)$ and $x(n) \in X(n)$ is any closed point near to x , then for the lexicographie ordering we have

$$(\dim D_x(X) , \delta(X,x)) > (\dim D_{x(n)}(X(n) , \delta(X(n),x(n)) .$$

If x lies on an equimultiple curve D and if again $x(n) \in X(n)$ is any closed point near to x , then

$$(\dim D_x(X) , \delta(X,x)) \geq (\dim D_{x(n)}(X(n)) , \delta(X(n),x(n))) ,$$

and this inequality will be strict if D has been blown up during the algorithm. This allows to assign an invariant to x , which will drop strictly (in the lexicographic order) in every step of our algorithm. Namely, we take as components of our invariant the following numbers (in the given order):

- number of components of dimension 1 of $Sam(X)$;
- $\sum_\xi \delta(X,\xi)$, where ξ are the generic points of one-dimensional components of $Sam(X)$ such that $\dim D_\xi(X) = 1$;
- $\sup\{\delta(X,)\,|\,x$ an isolated point of $Sam(X)$ such that $\dim D_x(X) = 1\}$;
- number of isolated points of $Sam(X)$ such that $\dim D_x(X) = 1$;
- number of isolated points of $Sam(X)$.

<u>REMARK.</u> As we can see, the algorithm is not canonical, since there is obviously a choice at each step.

<u>II Jung's method locally</u> (in characteristic 0)

In this paragraph we assume that our given scheme Z is a k-scheme for some algebraically closed field k of characteristic zero. In the first part of [18], Zariski describes an algorithm by which one can reach the following situation:

(3) At every closed point x of $\mathrm{Sam}(X)$ there is a transversal projection p : $\mathrm{Spec}\, k[[u_1,u_2]] \leftarrow \mathrm{Spec}\, \hat{O}_{X,x}$ with a normal crossing discriminant Δ such that $I(\Delta_{red}) \subset (u_1,u_2)$.

(where $\hat{O}_{X,x}$ denotes the completion of $O_{X,x}$, of course.)

PROPOSITION 4. If x is a closed point of $\mathrm{Sam}(X)$ satisfying (3), and if $\dim D_x(X) = 2$, then x is a quasi-ordinary point and (y,u_1,u_2) satisfies (2), where (u_1,u_2) is given by (3) and y is chosen such that $\Delta(f;u_1,u_2;y)$ is minimal.

Proof. Let ν be the multiplicity of X at x, let $N = \nu!$ and consider the ramified covering $S_N \to S$, where $S = \mathrm{Spec}\, k[[u_1,u_2]]$, $S_N = \mathrm{Spec}\, k[[x_1,x_2]]$ and $x_1^N = u_1$, $x_2^N = u_2$. After applying Tschirnhausen transformation to the equation f of X, f as an element of $\hat{O}_{Y,x} \otimes_S k[[x_1,x_2]]$ can be written as

$$f = \Pi(y-r_\alpha(x_1,x_2)),\quad r_\alpha = x_1^{m_1} x_2^{m_2} h_\alpha(x_1,x_2) \text{ and at least one } h_\alpha(0,0) \neq 0 .$$

Then I claim that $\Delta(f;u_1,u_2)$ has only one vertex of coordinates (λ_1,λ_2), where $\lambda_i = m_i/N$. First we note that

$$(4)\qquad \Delta(f;x_1,x_2;y) = (m_1,m_2) + \mathbb{R}_+^2$$

Now

$$f = y^\nu + \sum y^{\nu-i} a_{i,j,k}\, u_1^j u_2^k \in k[[y.u_1,u_2]]$$

$$= y^\nu + \sum y^{\nu-i} a_{i,j,k}\, x_1^{jN} x_2^{kN} \in k[[y,x_1,x_2]]$$

So, whenever $a_{i,j,k} \neq 0$, then $\left(\frac{Nj}{i}, \frac{Nk}{i}\right) \in \Delta(f;x_1,x_2;y)$ and therefore $(j/i,k/i) \in (\lambda_1,\lambda_2) + \mathbb{R}_+^2$, which implies

$$(5)\qquad \Delta(f;u_1,u_2;y) \subset (\lambda_1,\lambda_2) + \mathbb{R}_+^2 .$$

Since we have applied Tschirnhausen transformation to f , we know that $(f;u_1,u_2;y)$ is minimal. So to prove equality in (5) , it remains to show that $(\lambda_1,\lambda_2) \in (f;u_1,u_2;y)$. Suppose not. Then we know from [6] , Remarks on p. 27, that $\delta(f;u_1,u_2) > \lambda_1 + \lambda_2$. So whenever $a_{i,j,k} \neq 0$, then $j + k > i(\lambda_1 + \lambda_2)$ and consequently $jN + kN > i(m_1 + m_2)$. This would imply $\delta(f;x_1,x_2;y) > i(m_1 + m_2)$ in contradiction to (4) .

So we have seen that condition (3) implies the condition (*) of Proposition 3, and together with Propositions 1 and 2 this gives an algorithm for the desingularization of surfaces over an algebraically closed field of characteristic zero. Moreover we note that once condition (3) is satisfied, this algorithm is the same as Zariski's in [18] .

EXAMPLE. Let

$$f = y^3 + 3yu_1^4 u_2^4 + 2u_1^6u_2^6 + u_1^7u_2^7(u_1+u_2^2) \in \mathbb{C}[\![y,u_1,u_2]\!]$$

Then

$$\Delta(f;u_1,u_2) = (2,2) + \mathbb{R}_+^2 .$$

The discriminant defined by the projection $\mathbb{C}[\![u_1,u_2]\!] \hookrightarrow \mathbb{C}[\![y,u_1,u_2]\!]$ is

$$4(3u_1^4u_2^4)^3 - 27(4u_1^6u_2^6 + u_1^7u_2^7(u_1+u_2^2))^2$$

$$= -27u_1^{13}u_2^{13}(u_1+u_2^2)(4 + u_1u_2(u_1+u_2^2)) ,$$

and this discriminant does not have normal crossings. This shows that the converse of Proposition 4 is false in general.

III Abhyankar's proof

In this paragraph, X is a reduced surface embedded in an excellent regular scheme Z of dimension 3.

DEFINITION. We call bad points of X the closed points of $\mathrm{Sam}(X)$ satisfying one of the following conditions:

a) The point is isolated in $Sam(X)$.

b) The point is a singular point of a one-dimensional component of $Sam(X)$.

c) The point is in the intersection of one-dimensional components of $Sam(X)$.

d) If $X = X(0) \leftarrow X(1) \leftarrow \ldots \leftarrow X(n)$ is any sequence obtained by successively blowing up any regular equimultiple curve through the given points resp. near points to the given point, then $\nu(X(i)) = \nu(X)$ for all i .

Of course, a closed point of $Sam(X)$ which is not bad is called good.

PROPOSITION 5. There is only a finite number of bad points.

Proof. We have only to prove that the points satisfying d) are finite in number. Take any one-dimensional component Y of $Sam(X)$ with genric point y . We apply the algorithm of blowing up first Y and then closure of points near to y in the successive transforms of X . From the theory of desingularization of curves we know that this algorithm is finite. Let n be the last step and $X(n)$ the strict transform of X . Since $X(n)$ does not contain any point near to y , we see that above $Y = \overline{\{y\}}$ there is at most a finite number of points $x' \in X(n)$ such that $\nu(x') = \nu(X)$. The projection of these finitely many points may be bad points, but all other points of $Y\text{-}Reg(Sam(X))$ are good.

PROPOSITION 6. (Abhyankar) The following algorithm is finite: We blow up every bad point of X and do the same for all the successive transforms $X(n)$ of $X = X(0)$ as long as $\nu(X(n)) = \nu(X)$.

REMARK 6.1. Proposition 6 reduces the problem of resolution of the singularities of X to the case where all closed points of $Sam(X)$ are good. In that particular case the following algorithm will be finite: We blow up any curve of $Sam(X)$ and do the same for the successive transforms $X(i)$ of $X = X(0)$ as long as $\nu(X(i)) = \nu(X)$. Furthermore, if $X(n)$ is the last transform of X , then $\nu(X(n)) < \nu(X)$. The proof of this Remark is clear from the definition of good points.

We will give a proof of Proposition 6 based on our notion of a quasi-ordinary singularity, thereby linking it to Zariski's proof. The result will follow from two facts to be proven below:

a) By successively blowing up bad points with 2-dimensional

directrix, all of them will eventually become quasi-ordinary, provided that they remain bad (Lemma 6.2).

b) By successively blowing up quasi-ordinary bad points, all of them will eventually become good, provided that the multiplicity remains unchanged (Lemma 6.3). *)

LEMMA 6.2. If we apply the algorithm of Proposition 6 then, for all large n, if $x(n) \in X(n)$ is any bad point such that $\dim D_{x(n)}(X(n)) = 2$, $x(n)$ is quasi-ordinary.

For the proof we restrict ourselves to the most difficult case, namely that of prime characteristic p such that $\nu(X) \equiv 0(p)$.

REMARK 6.2.1. From [12], p.109 we know that for large n the one dimensional components of $\mathrm{Sam}(X(n))$ will be regular and will have normal crossings. Therefore, we make the additional assumption:

(**) The components of $\mathrm{Sam}(X)$ are regular and have normal crossings.

DEFINITION 6.2.2. Let $x \in X$ be a bad point such that $\dim D_x(X) = 2$ and assume that X satisfies (**). Let

$$X(n) \xrightarrow{\pi(n-1)} X(n-1) \longrightarrow \dots \xrightarrow{\pi(1)} X(1) \xrightarrow{\pi(0)} X(0) = X$$

be a sequence obtained by blowing up a point $x(i) \in X(i)$ near to x at each step $(x(0) = x)$. Let $E_{i,n}$ denote the strict transform in $X(n)$ of the exceptional divisor of $\pi(i)$. Let $Y(0)$ denote the one-dimensional components of $\mathrm{Sam}(X)$ (there are at most 2), and for $n \geq 1$ let $Y(n)$ denote the strict transform in $X(n)$ of the one-dimensional part of $\mathrm{Sam}(x)$ passing **through** $x(n)$. (Note that

*) Note from the editor: Here is a gap for bad points with one dimensional directrix, which may be filled in the following way: By Proposition 4 we know that the multiplicity of such a point x will strictly decrease by a suitable sequence of blowing up points and regular equimultiple curves. Now we use a result of Zariski (Reduction of the singularities of algebraic three-dimensional varieties, Ann. Math. 45 (1944), 472 - 542) which is called "dominant character of a normal sequence" and which says that as long as the multiplicity does not change, the order of blowing up a point and a curve may be interchanged. Therefore, it we first perform all blowing-ups centered at points, the point has become good since now multiplicity will be decreased by blowing up curves only.

either $Y(n) = \emptyset$ or $Y(n)$ is a regular, irreducible curve for $n \geq 1$.) Then we define

$$\omega(x(n)) = \delta(X(n),x(n)) - \sum_{Y(n)} \delta(X(n),Y(n)) - \sum_{i} [\delta(X(i),x(i)) - 1] \mathrm{ord}_{x(n)} E_{i,n}.$$

The picture below shows the characteristic polyhedron of $x(n)$ in the case that $n \geq 1$ and $Y(n) \neq \emptyset$. In this case we may choose a regular system of parameters (y,u_1,u_2) at $x(n)$ (in $Z(n)$, the transform of Z) , such that

$$\begin{cases} (u_1) = I(E_{n-1,n}) \;, \quad (y,u_2) = I(Y(n)) \text{ and} \\ \Delta(f;u_1,u_2;y) \text{ is minimal,} \end{cases}$$

f being the equation of $X(n)$ ([12] , p. 28; see also the proof of Lemma 6.2.3. below).

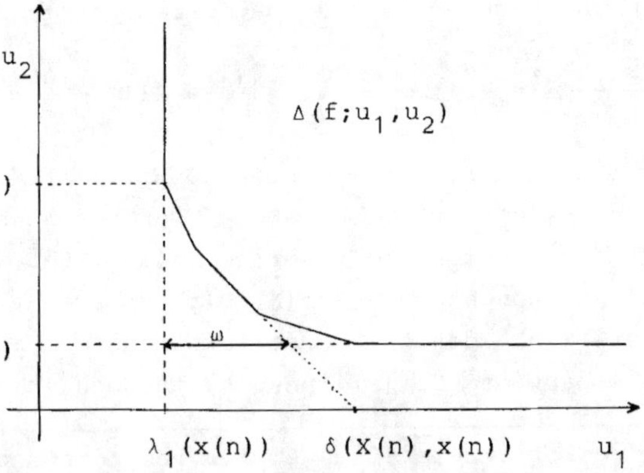

Here $\lambda_1(x(n)) = \delta(X(n-1),x(n-1))$, $\lambda_2(x(n)) = \delta(X(n),Y(n))$ and $\omega = \omega(x(n)) = \delta(X(n),x(n)) - \lambda_1(x(n)) - \lambda_2(x(n))$.

The proof of Lemma 6.2. will consist in a description of the number $\omega(x(k))$ under blowing up $x(k)$. To simplify notations, we restrict the computations to the case that there is at most one $E_{i,k}$ passing through $x(k)$, and we also assume that $k \geq 1$. It will become clear that the other cases may be treated in exactly the same way. So we may assume that we can choose a regular system of parameters (y,u_1,u_2) at $x(k)$ with the following properties ($Z(k)$ denotes the k-th transform of Z):

(6)

a) $\mathrm{in}_{x(k)}(y) = I(D_{x(k)}(X(k))) \subset gr_{x(k)} O_{Z(k),x(k)}$.

b) $\Delta(f;u_1,u_2;y)$ is minimal, when f is the equation of $X(k)$ in $Z(k)$.

c) If $\mathrm{ord}_{x(k)}(E_{i,k}) \neq 0$ for some i , then $I(E_{i,k}) = (u_1)$ and $\delta(X(i),x(i)) - 1 = \inf\{x_1 \mid (x_1,x_2) \in \Delta(f;u_1,u_2;y)\} = \lambda_1(x(i))$.

d) If $Y(k) \neq \emptyset$, then $I(Y(k)) = (y,u_2)$ and $\delta(X(k),Y(k)) = \inf\{x_2 \mid (x_1,x_2) \in \Delta(f;u_1,u_2)\} = \lambda_2(x(k))$.

(For $k = 0$, the existence of such y,u_1,u_2 is shown in [12] .) We put $\delta = \delta(X(k),x(k))$, $\lambda_1 = \lambda_1(x(i))$, $\lambda_2 = \lambda_2(x(k))$ and we note that then

(7) $$\omega(x(k)) = \delta - \lambda_1 - \lambda_2 .$$

($\lambda_1 = 0$ if there is no $E_{i,k}$.) $Z(k+1)$ is covered by three affine pieces. On the piece with exceptional divisor $y = 0$ there is no point near to $x(k)$, so we assume that $u_1 = 0$ is the equation of the exceptional divisor (again, the case $u_2 = 0$ is similar) .

Let K be the residue field at $x(k)$. Then at $x(k+1)$ we have a regular system of parameter $(y',u_1,\tilde{\Phi}(1,u_2'))$, where $y' = y/u_1$, $u_2' = u_2/u_1$ and where $\tilde{\Phi}(U_1,U_2)$ is the lifting of an irreducible homogeneous polynomial $\Phi(U_1,U_2) \in K[U_1,U_2]$ (see [6] , Construction of the parameters and Lemma 4).

Putting $\nu = \nu(X)$, we may write

$$f = y^\nu + \sum_{i=1}^{\nu} y^{\nu-1} f_i(u_1,u_2) .$$

Then

$$F = \mathrm{in}_\delta(f) = Y^\nu + \sum_{i=1}^{\nu} Y^{\nu-i} U_1^{\alpha(i)} U_2^{\beta(i)} Q_i(U_1,U_2) ,$$

where $Y = \mathrm{in}_\delta(y)$, $U_j = \mathrm{in}_\delta(u_j)$, and $Q_i = 0$ or Q_i is homogeneous of degree $i\delta - \alpha(i) - \beta(i)$. We assume that Q_i is divisible neither by U_1 nor by U_2 : Lifting Q_i to polynomials $\tilde{Q}_i$ with coefficients in $O_{Z(k),x(k)}$, we have

$$f' = f/u_1^{\nu} = y'^{\nu} + \sum_{i=1}^{\nu} y'^{\nu-1} u_1^{i(\delta-1)} [\tilde{Q}_i(1,u_2')u_2'^{\beta(i)} + u_1 h_i]$$

where $h_i \in 0_{Z(k),x(k)}[y/u_1, u_2/u_1]$ (see [6], Lemma 4). Now we define

$$\beta(f', u_1, \tilde{\Phi}(1,u_2'), y') = \inf\{\tfrac{1}{i}\mathrm{ord}_{x(k+1)} \tilde{Q}_i(1,u_2')u_2'^{\beta(i)} \mid Q_i \neq 0\}$$

<u>LEMMA 6.2.3.</u> With the notations introduced above, we have

a) $\omega(x(k+1)) \leq \omega(x(k)) + \frac{1}{p}$.

b) If $\delta(X(k), x(k)) \in \mathbb{N}$ and $\omega(x(k+1)) < 1$, and if $Y(k+1) = \emptyset$ and no $E_{i,k+1}$ passes through $x(k+1)$, then $x(k+1)$ is a good point.

c) If $x(k+1)$ is a bad point and not rational over $x(k)$, then

$$\omega(x(k+1)) \leq \omega(x(k)) .$$

d) The inequality of c) is strict except for the following cases:

$$\begin{cases} \omega(x(k)) = 0 & \text{or} \\ \omega(x(k)) = 1 , & \beta' = 1 \text{ and } \deg\tilde{\Phi} = p = 2 . \end{cases}$$

where $\beta' = \beta(f', u_1, \tilde{\Phi}(1,u_2'), z')$ for a suitable choice of z'.

Proof. If (y', u_1, u_2') is a system of parameters, then it has again the properties stated in (6), and therefore,

$$\omega(x(k+1) \leq \tfrac{1}{i}\mathrm{ord}_{x(k+1)} \tilde{Q}_i(1,u_2')u_2'^{\beta(i)} \quad \text{for every } i \text{ such that } Q_i \neq 0$$

and consequently

$$\omega(x(k+1) \leq \delta - \inf\{\tfrac{\alpha(i)}{i} \mid Q_i \neq 0\} - \lambda_2 \leq \delta - \lambda_1 - \lambda_2 = \omega(x(k)).$$

So to prove a) we may now assume that u_2' is invertible at $x(k+1)$. We write $\nu = p^s q$ with $(p,q) = 1$, and we consider the following conditions:

(A) $\delta = \delta(X(k), x(k)) \notin \mathbb{N}$,

(B) $Q_i \neq 0$ for some i, $1 \leq i \leq \nu - 1$,

(C) $Q_{p^s} = 0$,

(D) $\beta = \beta(f', u_1, \tilde{\Phi}(1, u_2'), y') \notin \mathbb{N}$.

The numbers $(\delta-1, \beta)$ are the coordinates of the vertex $v = v(f', u_1, \tilde{\Phi}(1, u_2'), y')$ of smallest abscissa of $\Delta(f'; u_1, \tilde{\Phi}(1, u_2'); y')$. I claim that if (A) , (B) , (C) or (D) is true, then v is not solvable ([9] , (3.8)). Indeed, if (A) or (D) holds, then the coordinates of v are not integer (compare [9] , (3.9.2)), and if (B) or (C) is satisfied then the initial form $\mathrm{in}_v(f')_{u_1, \tilde{\Phi}(1, u_2'), y'}$ is not a p^s-th power (compare [9] ,(3.8) and (3.9)). Therefore, if one of the conditions (A) - (D) is satisfied, then

$$\begin{aligned}(8) \qquad \omega(x(k+1)) &\leq \beta \leq \inf\{\tfrac{1}{i}\mathrm{ord}_{x(k+1)}\tilde{Q}_i(1, u_2') \mid Q_i \neq 0\} \\ &\leq \inf\{\deg Q_i / i \cdot \deg \Phi \mid Q_i \neq 0\} \\ &\leq (\delta - \tfrac{\alpha(i)}{i} - \tfrac{\beta(i)}{i})/\deg\Phi \leq \omega(x(k))/\deg\Phi .\end{aligned}$$

If moreover $\deg\Phi \geq 2$ ($\Leftrightarrow x(k+1)$ is not rational over $x(k)$) , we deduce

$$(9) \qquad \omega(x(k+1)) \leq \omega(k)/2 .$$

Now assume for a moment that $U_1^{\alpha(p^s)} U_2^{\beta(p^s)} Q_{p^s}(U_1, U_2) = P(U_1, U_2)^{p^s}$ for some $P \in K[U_1, U_2]$. If we change y to $z = \tilde{P}(u_1, u_2)$ ($\tilde{P}$ a lifting of P) , then the new parameters (z, u_1, u_2) at $x(k)$ still satisfy (6) , and putting $Z = \mathrm{in}_\delta z$ we see that the coefficient of $Z^{\nu-p^s}$ in $\mathrm{in}_\delta f'$ is zero, i.e. (C) is satisfied for the new parameters. Therefore, if none of the conditions (A) - (D) holds, we may assume (after a suitable choice of our coordinates not affecting (6)) that

$$(E) \quad \begin{cases} \delta \in \mathbb{N} \\ Q_i = 0 \quad \text{for all } i \text{ such that } p^s \nmid i ,\ 1 \leq i \leq \nu \\ Q_{p^s}(U_1, U_2) U_1^{\alpha(p^s)} U_2^{\beta(p^s)} \quad \text{is not a } p^s\text{-th power in } K[U_1, U_2] \end{cases}$$

If $v = v(f'; u_1, \Phi(1, u_2'); y')$ is still not solvable, then the inequalities (8) and (9) hold. If v is solvable, then we change y' to

$$z' = y' + a\,\tilde{\Phi}(1,u_2')^{\beta} u_1^{\delta-1} = y' + bu_1^{\delta-1}, a \in O_{Z(k+1),x(k+1)} ,$$

and we put $\beta' = \beta(f',u_1,\tilde{\Phi}(1,u_2'),z')$. If v' is the vertex of smallest abscissa of $\Delta(f';u_1,\tilde{\Phi}(1,u_2');z')$, then

$$v \in \Delta(f';u_1,\Phi(1,u_2')) \quad \text{(see [9] , (3.10))}$$

and v' has coordinates $(\delta-1,\beta')$ (see [6]). We denote by $\bar{b}$ the image of b in $O_{Z(k+1),x(k+1)}/(u_1,y') \cong K[U_2']_{(\Phi(1,U_2'))}$ (for this isomorphism se [6], Construction of the parameters). Then we get

$$\omega(x(k+1)) \leqq \beta' \leqq \mathrm{ord}_{x(k+1)} (\tilde{Q}_{p^s}(1,U_2')U_2'^{\beta(p^s)} + \bar{b}^{p^s})/p^s$$

$$\leqq \mathrm{ord}_{x(k+1)} (\partial(\tilde{Q}_{p^s}(1,U_2')U_2'^{\beta(p^s)})+1)/p^s ,$$

where $\partial \in \mathrm{Der}_{K^p[U_2'^p]}(K[U_2'])$. Now let r be maximal such that

$$U_1^{\alpha(p^s)} U_2^{\beta(p^s)} Q_{p^s}(U_1,U_2) = G^{p^r} \quad \text{for some } G .$$

Then $r < s$. If $r = 0$ then there exists $\partial \in \mathrm{Der}_{K^p[U_2'^p]}(K[U_2'])$ (either $\partial = \partial/\partial U_2'$ or ∂ is the extension of some $\partial' \in \mathrm{Der}_{K^p}(K)$) such that

$$\partial Q_{p^s}(1,U_2')U_2'^{\beta(p^s)} = \begin{cases} U_2'^{\beta(p^s)-1} \cdot H & \text{if } \partial = \partial/\partial U_2' \\ U_2^{\beta(p^s)} \cdot H & \text{otherwise.} \end{cases}$$

Here $H \in K[U_2'] \subset K[U_2']_{(\Phi(1,U_2'))}$ and $\deg_{U_2'}(H) \leqq \deg_{U_2'}(Q_{p^s})$. Then

$$\mathrm{ord}_{\Phi(1,U_2')} H \leqq \deg_{U_2'} Q_{p^s}(1,U_2') \leqq \deg Q_{p^s}(U_1,U_2) .$$

If $r > 0$ then a computation similar to the one given above yields

$$\mathrm{ord}_{x(k+1)} (H + \bar{b}^{p^{s-r}}) \leq \mathrm{ord}_{x(k+1)} (H) + 1 .$$

In any case we will have

$$\text{(10)} \quad \omega(x(k+1) \leq \deg Q_{p^s}/p^s \cdot \deg\Phi + 1/p^{s-r}$$

$$\leq \omega(x(k))/\deg\Phi + 1/p ,$$

and this proves a).

b). We have

(11) $\delta(X(k+1),x(k+1) = \delta - 1 + \omega(x(k+1)) = \text{integer} + \omega(x(k+1))$.

Let X'' be the blowing up of $X(k+1)$ at $V(y',u_1')$. Then X'' contains at most one point x'' which is near to $x(k+1)$, and the characteristic polyhedron of x'' (if it is near) will be $\Delta(f';u_1',\tilde{\Phi}(1,u_2');y')$ translated by $(-1,0)$ ([12] , p.127) . In particular $\delta(X'',x'') = \delta(X(k+1),x(k+1)) - 1$. Now x'' satisfies again the assumptions of (6) and the condition (11) (where the pair $(x(k),x(k+1))$ is replaced by $(x(k+1),x''))$. So the assertion of b) follows by descending induction on δ .

c) In the cases (A) - (D) , the assertion follows from (8) . If (E) is satisfied and $\omega(x(k+1)) < 1$, then $x(k+1)$ is good by b) . The remaining case is $\omega(x(k))/\deg\Phi + 1/p \geq \sup(\omega(x(k)),1)$, in which the result is clear.

If $\omega(x(k)) = \omega(x(k+1)) = 0$, then these points are quasi-ordinary. The case where $\omega(x(k+1)) = \beta' = 1$ will be studied in the next Lemma.

<u>LEMMA 6.2.4.</u> Assume that $x(k+1)$ is bad and very near to $x(k)$, and that $\delta(X(k),x(k) \in \mathbb{N}$. If, for some suitable choice of parameters $\beta(x(k+1)) = 1$, then $\omega(x(k+1)) = 1$. Let x'' be any point of $X(k+2)$ near to $x(k+1)$. Then we have :

a) If $x'' \in E_{k+1,k+2}$, then x'' is quasi-ordinary or $\dim D_{x''}(X(k+2)) = 1$.

b) If $x'' \notin E_{k+1,k+2}$, then either x'' is good, or x'' is rational over $x(k+1)$ and $\beta(x'') = \omega(x'') = 1$.

c) For i large enough, $x(k+i)$ is quasi-ordinary.

Proof. By Lemma 6.2.3 , b) , it is clear that $\omega(x(k+1) = 1$. Now given x'' , let $\Psi(1,\tilde{\Phi})$ be a parameter at x'' . If x'' is rational over $x(k+1)$, then by [6] , we have

$$\beta(f'/u_1^{2\nu},u_1,\Psi(1,\tilde{\Phi})) < \beta(f'/u_1^{\nu},u_1,\tilde{\Phi}(1,u_2')) .$$

Therefore x'' is good by Lemma 6.2.3 , b) . Assume now that $x'' \in E_{k+1,k+2}$ is very near to $x(k+1)$. Then

$$\omega(x'') \leq \delta - 1 + \beta(x(k+1)) - 1 - (\delta(X(k+1),x(k+1)) - 1)$$
$$= \delta - 1 + \beta(x(k+1)) - \delta(X(k+1),x(k+1)) = 0 ,$$

and therefore a) follows. In the situation of b) we know that

$$\beta\,(f'/u_1^{2\nu},u_1,\Psi(1,\tilde{\Phi})) \leq \beta(f'/u_1^{\nu},u_1,\tilde{\Phi}(1,u_2')) ,$$

and therefore either x'' is good or it satisfies the same hypothesis as $x(k+1)$. This proves b) and also c) , since the sequence of β's is not constant ([12]) .

LEMMA 6.2.5. With the hypotheses and notation of Lemma 6.2 , assume the sequence

$$x(k) \leftarrow x(k+1) \leftarrow \dots \leftarrow x(k+n) , n \geq 1$$

satisfies

$$\begin{cases} \omega(x(k)) > 0 , \\ \omega(x(k+1) \geq \omega((x(k)) \quad \text{for} \quad 0 \leq i \leq n , \\ x(k+i) \quad \text{is very near to} \quad x(k) \quad \text{and bad for} \quad 0 \leq i \leq n . \end{cases}$$

Then this sequence is finite.

Proof. If $x(k+i) \notin E_{k+i-1,k+i}$ for all i , $1 \leq i \leq n$, then Hironaka's proof in [12] together with [6] implies that the sequence $\omega(x(k+j))$ is constant by (8). From [12], we know that

$$\omega(x(k+i)) \leqq \beta(x(k+i-1)) - \omega(x(k+i-2) - \lambda_2(x(k+i-1)) ,$$

and by (8) we have

$$\beta(x(k+i-1)) \leqq \omega(x(k+i-2)) + \lambda_2(x(k+i-1)) ,$$

and therefore $\omega(x(k+i)) = 0 \leqq \omega(x(k))$. So we may now assume that (E) is satisfied at $x(k+j)$ for some $j < i$, and we choose j minimal for this property. Then $\beta(x(k+j+1)) \leqq \omega(x(k)) + 1/p$ by the proof of (10). If now $\omega(x(k)) \leqq 1/p$ then $\beta(x(k+j+1)) \leqq 2/p \leqq 1$, and the assertion follows from Lemma 6.2.4 in the case $\beta(x(k+j+1)) = 1$, while $\beta(x(k+j+1)) < 1$ implies that $x(k+j+1)$ is a good point. Assume finally that $\omega(x(k)) > 1/p$. Then the result follows from the inequality

$$\omega(x(k+1)) \leq \beta(x(k+i-1)) - \omega(x(k)) \leq \frac{1}{p}$$

([12] , together with Lemma 6.2.3).

In the following Lemma, we denote by $\{z\}$ the fractionary part of a real number z .

Lemma 6.3. With the notations of Proposition 1, assume that $x(k) \in X(k)$ is a quasi-ordinary bad point (with $\dim_{x(k)} X(k) = 2$), and let (λ_1, λ_2) be the vertex of the polyhedron $\Delta(f; u_1, u_2)$. Then

a) $\{\lambda_1\} + \{\lambda_2\} \geq 1$

b) After blowing up $x(k)$ there are at most two bad points very near to $x(k)$, and these points are quasi-ordinary. The coordinates of the vertices of the respective polyhedra are $(\lambda_1', \lambda_2') = (\lambda_1+\lambda_2-1, \lambda_2)$ and $(\lambda_1'', \lambda_2'') = (\lambda_1, \lambda_1+\lambda_2-1)$.

Proof: a) Assume the contrary, i.e. $\{\lambda_1\} + \{\lambda_2\} < 1$. Then one of the λ's must be 1 since $\delta(X(k), x(k)) > 1$. Therefore one of $V(y, u_1)$ and $V(y, u_2)$ is permissible, and this is the only permissible curve through $x(k)$ by Proposition 3. If we blow up this curve, Proposition 1 tells us that there is at most one point very near to $x(k)$. This point is again quasi-ordinary and $\{\lambda_1\}, \{\lambda_2\}$ will not change, whereas δ drops. So we get the result by descending induction on δ .

b) Assume first that $\{\lambda_1\} + \{\lambda_2\} > 1$. Then $\delta(X(k), x(k)) = \lambda_1 + \lambda_2 \notin \mathbb{N}$ (case (A) in Lemma 6.2.3) , and therefore ω is zero at every point very near to $x(k)$. If u_2/u_1 is invertible at such a point, then the point is good by a) . The remaining points have local coordinates $(y/u_1, u_1, u_2/u_1)$ or $(y/u_2, u_1/u_2, u_2)$, and for these the assertion is clear. By a) it remains to consider the case $\{\lambda_1\} + \{\lambda_2\} = 1$, which implies $\lambda_1 + \lambda_2 = \delta(X(k), x(k)) \in \mathbb{N}$. Let x' be a point very near to $x(k)$. If u_2/u_1 is invertible at x' , then $\omega(x') \leq \omega(x(k)) + \frac{1}{p} = \frac{1}{p}$ by (10) . But this means that x' is good, since it satisfies the hypotheses of part b) of Lemma 6.2.3. If x' has local parameters $(y/u_1, u_1, u_2/u_1)$ or $(y/u_2, u_1/u_2, u_2)$, then it is again quasi-ordinary, and so by Lemma 6.2.3., b) it must even be a good point.

CONCLUSION. Using the notation of Lemma 6.3 we have

$$\{\lambda_1'\} + \{\lambda_2'\} < \{\lambda_1\} + \{\lambda_2\} \quad \text{and} \quad \{\lambda_1''\} + \{\lambda_2''\} < \{\lambda_1\} + \{\lambda_2\}$$

Therefore, by Lemma 6.2 and 6.3, the algorithm of 6.1 will lead to a situation where $\dim_x X \leq 1$ for every bad point x of X. In this case, the proof of Proposition 6 follows from [12], p. 114.

IV Hironaka's method

We will use the assumptions and notations of III. Hironaka answers the question by looking at the deformations of the characteristic polyhedra. There are three steps in his proof (given in [6] and [12]):

Step 1. We can reduce the problem to the case where all points in Sam(X) are isolated ([12], p.109).

Step 2. The difficult case is the one in which the dimension of the directrix of X at x is 2.

Step 3. Let (α,β) be the vertex of smallest abscissa of $\Delta(f;u_1,u_2)$ and choose u_1,u_2 so that β is minimal. Then β will drop along the sequence

$$X = X(0) \leftarrow X(1) \leftarrow \dots \leftarrow X(n) \leftarrow \dots ,$$

where $X(n)$ is obtained from $X(n-1)$ by blowing up $\mathrm{Sam}(X(n-1))$.

REFERENCES.

[1] S. ABHYANKAR, Local uniformization on algebraic surfaces over ground fields of characteristic $p \neq 0$. Ann. of Maths., 63 (1956), p. 491-526.

[2] S. ABHYANKAR, On the ramification of algebraic functions, Ann. of Maths., 77 (1955), p. 575-592.

[3] S. ABHYANKAR, Reduction to multiplicity less than p in a p-cyclic extension of a two dimensional regular ring, Math. Ann. 154 (1964), 28-55

[4] S. ABHYANKAR, Nonsplitting of valuations in extensions of two dimensional regular local domains, Math. Ann. 170 (1967), 87-144.

[5] V. COSSART, Une nouvelle définition de l'invariant δ , Unpublished.

[6] V. COSSART, Desingularization of embedded excellent surfaces. Tohoku Math. Journ. 33 (1981), 25-33.

[7] J. GIRAUD, Etude locale des singularités, Orsay, Pub. no. 26 (1972)

[8] J. GIRAUD, Desingularization in low dimension, Lecture 2 in this volume.

[9] H. HIRONAKA, Characteristic polyhedra of singularities, J. Math. Kyoto University, V. 7 (1967), 251-293.

[10] H. HIRONAKA, Bimeromorphic smoothing of a complex analytic space, University of Warwick (1967).

[11] H. HIRONAKA, Introduction to the theory of infinitely near singular points, Memorias de Matematica del Instituto "Jorge Juan" 28.

[12] H. HIRONAKA, Desingularization of excellent surfaces. Advanced Science Seminar in Algebraic Geometry (summer 1967). Notes by B.Bennett, Bowdoin College, reprinted in the appendix of this volume.

[13] H. HIRONAKA, J.M. AROCA, J.U. VICENTE, The theory of the maximal contact. Memorias de Matematica del Instituto "Jorge Juan" 29.

[14] H. HIRONAKA, Desingularization Theorems. Memorias de Matematica des Instituto "Jorge Juan" 30.

[15] H.W.E. JUNG, Darstellung der Funktionen eines algebraischen Körpers zweier unabhängigen Veränderlichen in der Umgebung einer Stelle. J. Reine Angew. Math. 133 (1908), 289-314.

[16] J. LIPMAN, Desingularization of two-dimensional schemes. Ann. of Math. 107 (1978).

[17] T. SANCHEZ, Teoria de singularidades de superficies algebroides sumergidas. Monografias y memorias de Matematica. IX, Pub del Instituto "Jorge Juan" de Matematicas. Madrid (1976).

[18] O. ZARISKI, Exceptional singularities of an algebroid surface and their reduction. Atti. Acad. Naz. dei Lincei, serie VIII, vol. XLIII (1967), 135-146.

A P P E N D I X

Desingularization of Excellent Surfaces

Heisuke Hironaka
Advanced Science Seminar in Algebraic Geometry
Bowdoin College, Summer 1967

Notes by Bruce Bennett

I. Fundamental Concepts

1. Excellence

Def. A scheme X is excellent if

(1) X is noetherian

(2) $\forall : X' \longrightarrow X$ of finite type, $\mathrm{Sing}(X')$ is closed

($\mathrm{Sing}(X) = \{x \in X \mid \mathcal{O}_{X,x}$ is not a regular local ring$\}$)

(3) $\forall X'' \xrightarrow{f} X' \xrightarrow{g} X$ both of finite type,

$\forall x' \in X'$, if $R = \hat{\mathcal{O}}_{X',x'}$, let $\tilde{X}'' = \mathrm{Spec}\, R \times_{X'} X''$.

(deduced from $\mathrm{Spec}\, R \longrightarrow X'$ via base extension by f)

and let h be the projection $h: \tilde{X}'' \longrightarrow X''$.

Thus:

$$\begin{array}{ccc} X'' \times_{X'} \mathrm{Spec}\, R = \tilde{X}'' & \xrightarrow{h} & X'' \\ \downarrow & & \downarrow f \\ \mathrm{Spec}\, R & \longrightarrow & X \end{array}$$

Then h has the property: $\mathrm{Sing}(\tilde{X}'') = h^{-1}(\mathrm{Sing}(X''))$.

Remarks: Every scheme of finite type over an excellent scheme is excellent, and in particular

Every closed subscheme of an excellent scheme is excellent.

Spec (any complete local ring) is excellent (a theorem of Nagata and Grothendieck, E.G.A.IV).

Any finite type scheme over $\mathbb{Z}$ (= integers) is excellent (Nagata).

2. Tangent Spaces and Cones

X a scheme, $x \in X$. $\dim_x(X)$ = Krull dim $\mathcal{O}_{X,x}$, $\dim X = \underset{x \in X}{\mathrm{Max}}(\dim_x(X))$

Let $\mathfrak{m} = \mathfrak{m}_{X,x}$, $\mathcal{O} = \mathcal{O}_{X,x}$, $\varkappa(x) = \mathcal{O}/\mathfrak{m}$, and define:

$T_x(X)$ = Zariski tangent space to X at x = Spec $(S_{\varkappa(x)}(\mathfrak{m}/\mathfrak{m}^2))$.

($S_{\varkappa(x)}(-)$ denotes symmetric algebra over $\varkappa(x)$).

$C_x(X)$ = Tangential cone of X at x = Spec $(\mathrm{Gr}_x(X))$

($\mathrm{Gr}_x X = \mathrm{Gr}_{\mathfrak{m}}(\mathcal{O})$).

Note: $C_{x(x)} \hookrightarrow T_x(x)$ via the canonical surjection

$S_{\varkappa(x)}(\mathfrak{m}/\mathfrak{m}^2) \longrightarrow \mathrm{Gr}_{\mathfrak{m}}(\mathcal{O}) \longrightarrow 0.$

$\mathcal{T}_x(X)$ = Strict Tangent Space of X and x = **maximum** linear subspace T of $T_x(X)$ (passing through origin) such that equivalently:

(i) $C_x(X) = C_x(X) + T$ (+ denotes addition of points in

$T_x(X) = \mathbb{A}^n_{\varkappa(x)}$, $n = \dim_{\varkappa(x)} \mathfrak{m}/\mathfrak{m}^2$).

(ii) $C_x(X) \hookleftarrow T$, and $C_x(X) \cong T \times S$, where

$S \hookrightarrow C_x(X)$ is some closed subspace.

Remark: The existence of $\mathcal{T}_x(X)$ is proved by showing that if two linear subspaces T_1 and T_2 of $T_x(X)$ have property (i) or (ii) , then so does $T_1 + T_2$.

Observe x is a regular point of X $\Longleftrightarrow$ $T_x(X) = C_x(X) = \mathcal{T}_x(X)$.

Now suppose $x \in X \hookrightarrow Z$. $R = \mathcal{O}_{Z,x}$, $\mathcal{O} = \mathcal{O}_{X,x}$,

M (resp. $\mathfrak{m}$) = maximal ideal of R (resp $\mathcal{O}$). $k = R/M = \mathcal{O}/\mathfrak{m}$

J = ideal of X in Z at x, so that we have

$$0 \longrightarrow J \longrightarrow R \longrightarrow \mathcal{O} \longrightarrow 0 .$$

We obtain:

$$\begin{array}{ccccccccc} 0 & \longrightarrow & Gr_x(X,Z) & \longrightarrow & Gr_x(Z) & \longrightarrow & Gr_x(X) & \longrightarrow & 0 \\ & & \| & & \| & & \| & & \\ 0 & \longrightarrow & In_M(J,R) & \longrightarrow & \bigoplus_{\nu=0}^{\infty} M^{\nu}/M^{\nu+1} & \longrightarrow & \bigoplus_{\nu=0}^{\infty} \mathfrak{m}^{\nu}/\mathfrak{m}^{\nu+1} & \longrightarrow & 0 \end{array}$$

where $In_M(J,R)$ may be described as

(i) the ideal of $Gr_M(R)$ $(= Gr_x(Z))$ generated by all $In_M(f)$, $f \in J$, where $In_M(f) = f(\mathrm{mod}\ M^{\nu_M(f)})$,

where $\nu_M(f) = \begin{cases} \text{highest power of } M \text{ containing } f & \text{if } f \neq 0 \\ \infty & \text{if } f = 0 \end{cases}$

or (ii) the homogeneous ideal of $Gr_M(R)$ whose ν^{th} piece is:

$$J \cap M^{\nu} + M^{\nu+1} / M^{\nu+1}$$

In this situation we may describe $\mathcal{T}_x(X)$ as follows:

$\exists$ <u>minimum</u> k submodule T of $Gr^1_M(R)$ such that(if k[T] denotes $S_k(T) \hookrightarrow Gr_M(R)$),

$$(Gr_x(X,Z) \cap k[T])\, Gr_x(Z) = Gr_x(X,Z)$$

(i.e. $In_M(J,R)$ may be generated by Forms in k[T]).

Then $\mathcal{T}_x(X) \cong \mathrm{Spec}\ (Gr_x(Z)/_{T\ Gr_x(Z)})$ in the sense that:

$$\mathcal{T}_x(X) \hookrightarrow C_x(X) \hookrightarrow T_x(X) \hookrightarrow T_x(Z) .$$

(Note that $Gr_x(Z)$ is a polynomial ring since R is regular. k[T] is a polynomial subring and $\dim \mathcal{T}_x(X) = \dim T_x(Z) - \dim_k T$)

<u>Example</u>: If $Z = \mathbb{A}^2_k$, $X \hookrightarrow Z$ a curve, then $\mathcal{T}_x(X)$ = a point (as in the case of a node), or a line (as in the case of a cusp). More precisely, if X is locally defined in $\mathbb{A}^2_k (k = \bar{k})$ by f, where x = (0,0), write f as a power series

$$f = f_{\nu} + f_{\nu+1} + \ldots\ldots \quad (\nu = \mathrm{ord}\ f)$$

where $f_{\nu} = \prod_{i=1}^{\nu} (\alpha_i y - \beta_i x)$ $\quad \alpha_i, \beta_i \in k$. Then $\mathcal{T}_x(X)$ is a line <==> all the ratios $\alpha_i | \beta_i$ are the same. (In fact, the initial form of f at the origin is f_ν).

Terminology. $x \in X$, a scheme. x is a regular point (or a non-singular point or a simple point) of X if $O_{X,x}$ is a regular local ring. Otherwise, x is called a singular point of X.

3. Normal Flatness

X an excellent scheme $D \hookrightarrow X$ closed, reduced, irreducible subscheme. Define

$N(D,X)$ = The Normal cone of X along D = a family of cones parametrized by D = Spec $(Gr_D(X))$.

where $Gr_D(X) = \bigoplus_{\nu=0}^{\infty} I_D^\nu / I_D^{\nu+1}$ (I_D = ideal of D in X),

viewed as a sheaf of graded O_D- algebras.

Definition: $X \hookleftarrow D \ni x$. X is normally flat along D at x if D is non-singular and $Gr_D^\nu(X)$ is locally free at x $\forall \nu$.

X is normally flat along D if it is so at every point of D.

Note: It follows from general facts about flatness and the definition of excellence that $\{x \in D | X$ is normally flat along D at $x\}$ is open in D.

$\forall x \in D$, can define $N_x(D,X) = N(D,X) \times_D \text{Spec } \varkappa(x)$ = the fibre above x in the fibre system $N(D,X)$.

Theorem. $x \in D \hookrightarrow X$, D nonsingular. Then the following are equivalent:

1) X is normally flat along D at x.

2) $C_x(X) = T_x(D) \times N_x(D,X)$ (i.e., $C_x(X)/T_x(D) = N_x(D,X)$) and $T_x(D) \hookrightarrow \mathcal{T}_x(X)$

Note that we always have canonical maps

$$T_x(D) \longrightarrow C_x(X) \longrightarrow N_x(D,X)$$

corresponding to:

$$\mathrm{Gr}_{\mathfrak{m}/I}(\mathcal{O}/I) \xleftarrow{\text{surjective}} \mathrm{Gr}_{\mathfrak{m}}(\mathcal{O}) \longleftarrow \mathrm{Gr}_I(\mathcal{O}) \otimes_{\mathcal{O}} k$$

3) If we suppose X is embedded in a non-singular excellent scheme Z, then 1) and 2) are equivalent to:

$\exists$ hypersurfaces $H_1, \ldots, H_m$ in Z (locally at x) such that

a) $H_i \hookleftarrow X$

b) $\bigcap_i C_x(H_i) = C_x(X)$

($\Longrightarrow \bigcap_i H_i = X$ at x, <u>strict</u> implication).

c) $\forall$ i, H_i has the same multiplicity at x as it has at the generic point of D, (where both are regarded as points of Z).

c) means: if f_i defines H_i, $(R,M) = \mathcal{O}_{Z,x}$, $P \hookrightarrow R$ defines D at x (P a regular prime in R), then $\nu_M(f_i) = \nu_P(f_i)$ $\forall$ i.

4. Monoidal Transformations

X a scheme, $D \hookrightarrow X$ closed subscheme, $I \hookrightarrow \mathcal{O}_X$ the ideal sheaf of D on X. Let $X' = \underline{\mathrm{Proj}}\ (\bigoplus_{\nu=0}^{\infty} I^\nu)$, where $\bigoplus_{\nu=0}^{\infty} I^\nu$ is viewed as a sheaf of graded $\mathcal{O}_X$-algebras. There is a natural projection $h: X' \longrightarrow X$ induced by $\mathcal{O}_X \xrightarrow{\sim} I^0 \hookrightarrow \oplus I^\nu$. <u>(X', h) is called the monoidal transform (or "blowing up") of X with center D.</u>

Remarks:

1) Let spec $A \hookrightarrow X$ be an open affine, $(g_0, \ldots, g_n)$ a set of generators of $I|\mathrm{Spec}\ A$. Then above Spec A, X' may be described as

$$\bigcup_{i=0}^{n} \mathrm{Spec}\ (A[g_0/g_i, \ldots, g_n/g_i])$$

where $A[g_0/g_i,\ldots,g_n/g_i]$ is viewed as an A-subalgebra of A_{g_i}, and h is induced by the natural maps

$$A \longrightarrow A[g_0/g_i,\ldots,g_n/g_i] .$$

2) $I\mathcal{O}_{X'}$ is an invertible sheaf, generated on an affine of type Spec $(A[g_0/g_i,\ldots,g_n/g_i])$ (over Spec $A \hookrightarrow X$) by g_i. Moreover (X',h) is universal attracting for X-schemes Y $Y \xrightarrow{f} X$ s t $I\mathcal{O}_Y$ is an invertible sheaf.

3) $h^{-1}(D) = X' \times_X D = \underline{\mathrm{Proj}}\ (\bigoplus_{\nu=0}^{\infty} I^{\nu}/I^{\nu+1}) = P(N(D,X))$.

where we use the notation:

cone = Spec (graded algebra)

P(cone) = Proj (same graded algebra).

Similarly, if $x \in D$, $h^{-1}(x) = P(N_x(D,X))$.

It is easily shown that h induces an isomorphism $X'-h^{-1}(D) \xrightarrow[\sim]{h} X-D$.

4) If $D = \{x\}$, a point, $h^{-1}(x) = P(C_x(X)) = P(N_x(x,X))$. Therefore if x is non-singular, $h^{-1}(x) = P(T_x(X)) = \mathbb{P}^{m-1}_{\kappa(x)}$ where $m = \dim_x(X)$.

5) Suppose $D \hookrightarrow X \hookrightarrow Z$ closed subschemes, and consider the monoidal transform $g: Z' \longrightarrow Z$ of Z with center D. The strict transform of X by g is by definition the smallest closed subscheme $X' \longrightarrow Z'$ which induces $g^{-1}(X-D)$ on $Z'-g^{-1}(D)$. (i.e. such that $X'-g^{-1}(D) \xrightarrow{g} X - D$). Then if $h = g|X'$, $h: X' \longrightarrow X$ is the monoidal transform of X with center D.

5. Statement of the Problem.

Resolution of singularities of excellent schemes may be stated in the form of the following

<u>Conjecture</u>: X an excellent scheme. Then $\exists$ a finite succession of monoidal transformations

$$X_n \xrightarrow{f_{n-1}} X_{n-1} \longrightarrow \cdots \longrightarrow X_1 \xrightarrow{f_0} X_0 = X$$

such that ① The center of f_i is $D_i \hookrightarrow X_i$, where X_i is normally flat along D_i.

② X_n is empty (which is a slick way of saying that $(X_{n-1})_{red}$ is non-singular).

The <u>Conjecture</u> is known when 1) X is char. 0 (i.e. when char $\kappa(x) = 0 \ \forall\ x \in X$) and dim X is arbitrary (See Hironaka's Annals paper of March 1964), and 2) dim $X \leq 2$ and char X is arbitrary. The latter fact is the subject of the present lectures, and we give a proof in the special case when X is assumed to be embedded in a non-singular excellent scheme Z of dimension 3, in other words, when X is locally a hypersurface in Z. We suppose in addition that Z is of finite type over $k = \bar{k}$.

These additional hypotheses permit a considerable lightening of the terminology (in particular the "measure" of the badness of a singularity becomes very simple) and the necessity of introducing technicalities to solve problems of a purely algebraic nature is avoided. The essential character of the proof remains unchanged, however, and in fact should be more clearly evident.

No reference for the general proof is available at the present time, but it probably will be published shortly.

We will need to assume the <u>resolution of singularities of excellent curves</u> (i.e. excellent schemes of dimension 1) in the form of the <u>Conjecture</u>.

II. Proof of Resolution for Surfaces (in a Special Case)

1. Basic structure of the Proof.

The situation is: Z a 3-dimensional non-singular algebraic scheme over $k = \bar{k}$, $X \hookrightarrow Z$ a reduced surface. We want to resolve the singularities of X in the sense of the Conjecture.

Let $x \in X$. $R = \mathcal{O}_{Z,x}$, M = maximal ideal of R, (f)R the ideal of X in Z at x. Define $\nu_x(X)$ = the multiplicity of X at x = $\nu_M(f)$ = order of f at x.

Note that x is non-singular $\Longleftrightarrow \nu_x(X) = 1 \Longleftrightarrow$ f is a regular parameter in R. Let

$$\nu = \nu(X) = \max_{x \in X} \{\nu_x(X)\}$$

Remark. In general, if μ is any integer, define

$$\mathrm{Sing}_{\mu}(X) = \{x \in X \mid \nu_x(X) \geq \mu\}$$

$$\mathrm{Sing}_{*}(X) = \mathrm{Sing}_{\nu}(X) = \{x \in X \mid \nu_x(X) = \nu\}$$

="maximal singular locus"

It is obvious that $\mathrm{Sing}_{\mu}(X) \supseteq \mathrm{Sing}_{\mu+1}(X)$, and we have the following general fact:

X an excellent scheme, μ an integer, then $\mathrm{Sing}_{\mu}(X)$ is a proper, closed subscheme of X. (Here $\nu_x(X)$, the multiplicity of $\mathcal{O}_{X,x}$, cannot in general be interpreted as the order of a single element. The above fact is just the assertion of the "upper semicontinuity" of multiplicity of local rings of points on an excellent scheme).

In our case we can sketch a particularly simple proof of the closedness of $\mathrm{Sing}_{\mu}(X)$. Namely if char k = 0, $\nu_x(X) = \mu \Longleftrightarrow \exists$ a (pure) differential operator D on Z, $D = \prod_i (\partial/\partial x_i)^{a_i}$ of order $\sum a_i = \mu$, and $(Df)_x$ is a unit, hence also a unit in some neighborhood U of x, so $\nu_y(X) \leq \mu \ \forall \ y \in U$. (If char. k = p, use a "Hasse differentiation" instead of an ordinary one)[1].

Now consider a non-singular closed subscheme $D \hookrightarrow Z$ contained in $\mathrm{Sing}_\nu(X)$. Thus D is a point or a non-singular curve lying in X. Apply the monoidal transformations with center D:

$$\begin{array}{ccccc} & & X' & \hookrightarrow & Z' \\ & & \downarrow h & & \downarrow g \\ D & \hookrightarrow & X & \hookrightarrow & Z \end{array}$$

In general, a monoidal tranformation is said to be permissible if the scheme in question is normally flat along the center of the transformation. In our case (since X is a hypersurface in Z, and all the points of D have the same multiplicity as points of X - namely ν) condition (3) of the theorem on Normal Flatness, p.(4), implies that h is permissible. Therefore

$$T_x(D) \hookrightarrow \mathcal{T}_x(X) \ (\hookrightarrow T_x(Z))$$

In fact, $T_x(D)$ is a linear factor of $C_x(X)$ (by condition (2) of the above cited theorem, which asserts: X normally flat along $D \Longleftrightarrow C_x(X) \xrightarrow{\sim} N_x(X,D) \times T_x(D)$), and $\mathcal{T}_x(X)$ is the largest linear factor.

The following fact, essentially due to Zariski, and which we will call Idea A, is of crucial importance in the resolution. It states:

---

[1]. Differentials need not play a role in this type of result. See for example the much more general result in Hironaka's Annals paper: Thm. 1, Chap. III {3}, p.218, which uses entirely different methods.

Let X be an excellent scheme, $D \hookrightarrow X$ closed subscheme, $h: X' \longrightarrow X$ the monoidal transform with center D, permissible. Let $x \in D$, $x' \in X'$ such that $h(x') = x$. Then the singularity of x' is no worse than the singularity of x, and if the singularity of x' is "as bad as the singularity of x", then x' is contained in $P(\mathcal{T}_x(X)/T_x(D))$.

Remarks. 1) We will not make precise at this time the general meaning of "as bad as" and "no worse than". However, in our situation, they mean $\nu_{x'}(X') = \nu_x(X)$ and $\nu_{x'}(X') \leq \nu_x(X)$ respectively. Thus one consequence of Idea A is $\nu(X') \leq \nu(X)$.

2) $P(\mathcal{T}_x(X)/T_x(D))$ is viewed as a closed subscheme of $P(N_x(D,X))$ ($= h^{-1}(x)$ by Remark (3), p.5). In fact, $P(\mathcal{T}_x(X)/T_x(D))$ makes sense since $T_x(D) \hookrightarrow \mathcal{T}_x(X)$ as we have just seen. Moreover

$$C_x(X) = C \times \mathcal{T}_x(X) \overset{\text{by normal}}{\underset{\text{flatness}}{=\!=\!=}} N_x(X,D) \times T_x(D)$$

$$\Longrightarrow C \times (\mathcal{T}_x(X)/T_x(D)) = N_x(X,D) ,$$

so can view $P(\mathcal{T}_x(X)/T_x(D)) \hookrightarrow N_x(X,D)$.

3) A proof of Idea A in our special case will be given in an appendix.

We can now list some immediate consequences of Idea A in our situation $h: X' \longrightarrow X$, center $D \hookrightarrow \mathrm{Sing}_\nu(X)$, h permissible. (Terminology: An $x \in \mathrm{Sing}_\nu(X)$ is called a ν-fold point. A curve contained in $\mathrm{Sing}_\nu(X)$ is called a ν-fold curve. Let $x \in D$, a ν-fold point. $e = \dim \mathcal{T}_x(X)$, $d = \dim D$. Note that $0 \leq d \leq 1$, and $0 \leq e \leq 2$ (since the smallest vector space T such that $\mathrm{In}_M(f) \in k[T]$ must have at least dimension 1, and $\dim \mathcal{T}_x(X) = 3 - \dim T$ in our case. See page (3)). Note also that since $T_x(D) \hookrightarrow \mathcal{T}_x(X)$, $d \leq e$.

Case 1. d = 1. Then e = 1 or 2

a) e = 1. Then $\dim \mathcal{T}_x(X)/_{T_x(D)} = 0$, so

$P(\mathcal{T}_x(X)/T_x(D))$ is empty, so there is no ν-fold point in X' lying above x (by Idea A).

b) e = 2. One can have at most one ν-fold point x' lying above x, and if there is, it is the point $P(\mathcal{T}_x(X)/T_x(D))$.

Case 2. d = 0. Then e = 0, 1, or 2.

a) e = 0. No ν-fold points above x.

b) e = 1 at most one ν-fold point above x, namely

$$P(\mathcal{T}_x(X)|T_x(D)).$$

c) e = 2 Then either there exists a finite number of ν-fold points above x (possibly none), or if there exists a ν-fold curve, it must be precisely $P(\mathcal{T}_x(X)|T_x(D))$, which is a line (in $h^{-1}(x)$), isomorphic to $\mathbb{P}^1_{\kappa(x)}$.

Notice that if ∃ a ν-fold curve E' on X', mapping onto D by h, then E' is unique and non-singular. Namely, if D is a point, E' must be a specific $\mathbb{P}^1$ by Case 2c). And if D is a curve then E' must be isomorphic to D by h, i.e., a uniquely possible point in the fibre of every point of D, (by case 1b).

We now give an outline of the Resolution procedure:

$$X^2 \hookrightarrow Z^3 \text{ non-singular.} \quad \nu = \nu(X).$$

Step I. If $\mathrm{Sing}_\nu(X)$ contains a singular irreducible curve, apply quadratic transformations to each of its singular points. We end up with: $\mathrm{Sing}_\nu(X)$ does not contain any singular irreducible curve.

Here we use the resolution of singularities of curves by monoidal transformations (necessarily quadratic). We of course view the transform of the curve as contained in the transform of X. After a finite number of steps we resolve the singularities of the curve. Need to note that <u>we can create no new irreducible singular ν-fold curves in this process.</u> In fact, any ν-fold curve lying above a blown up point must be a certain $\mathbb{P}^1$, as we have seen; and if this $\mathbb{P}^1$ is not ν-fold, any ν-fold point x' lying above x is either isolated in $\mathrm{Sing}_\nu(X')$, or lies on the strict transform D' of some curve D in $\mathrm{Sing}_\nu(X)$ (and it is well known that x' is not a singularity of D' if x is not a singularity of D).

<u>Note</u>: A quadratic transformation is <u>always</u> permissible.

<u>Step II</u>. If $\exists$ a non-singular irreducible curve Γ in $\mathrm{Sing}_\nu(X)$, apply monoidal transformation with center Γ. By induction applied to Spec (local ring at the generic point P of Γ in X) we come to: <u>$\mathrm{Sing}_\nu(X)$ is a finite number of points.</u>

The point is that Spec $(\mathcal{O}_{X,P})$ has dimension one, so can resolve it by permissible monoidal transformations, which we may view as base extensions of transformations beginning with X. Thus:

$$\begin{array}{ccc}
\text{permissible} \downarrow & & \downarrow \text{permissible} \\
(\mathrm{Spec}\ \mathcal{O}_{X,P})' & \longrightarrow & X' \\
\text{blow up closed point} \Big\downarrow & & \Big\downarrow \text{blow up } \Gamma \text{ permissible since } \Gamma \hookrightarrow \mathrm{Sing}_\nu(X) \\
\mathrm{Spec}\ \mathcal{O}_{X,P} & \longrightarrow & X
\end{array}$$

In this context, to say that we can resolve the singularities of Spec $\mathcal{O}_{X,P}$ means that eventually we obtain a curve $\bar{\Gamma}$ above Γ which has non-singular generic points (in particular, points of multiplicity 1), so by the upper semicontinuity of multiplicity, only a finite number of points of $\overline{\Gamma}$ can be ν-fold.

Step III. After eliminating ν-fold curves as in Step II, apply quadratic transformations to each point in $\mathrm{Sing}_{\nu}(X)$. By doing this we may create a new ν-fold curve. If so, apply Step II, and if not apply Step III again. Repeat.

We thus obtain a sequence of permissible monoidal transformations:

$$\longrightarrow X_{(m)} \xrightarrow{h(m-1)} X_{(m-1)} \longrightarrow \cdots\cdots \longrightarrow X_{(1)} \xrightarrow{h=h_{(0)}} X$$

Then: Theorem. For some m, $\nu(X_{(m)}) < \nu(X) = \nu$.

The proof of this theorem is of course the central problem. To prove it, we may start with a situation in which there are only isolated ν-fold points (i.e. start with Step III), in other words assume the ν-fold locus is discrete. The fact that ν-fold curves may be subsequently created is not crucial. The proof is done by contradiction.

We will start with: $x \in X$, isolated in $\mathrm{Sing}_{\nu}(X)$, and distinguish two cases: $e = 1$ and $e = 2$. ($e = \dim \mathcal{T}_x(X)$). Note that the case $e = 0$ is trivial, for then, by Idea A, there can be no ν-fold points above x. The case $e = 2$ is more difficult, this is the "cusp-type situation" in which 2-dimensional phenomena seem to be especially exhibited.

2. The Case $e = 1$.

Suppose we have an infinite sequence of permissible monoidal tranformations:

$$\begin{array}{ccccc} \longrightarrow \; . \; . \; . \; . \; . \; \longrightarrow & Z_1 & \xrightarrow{g} & Z^{(3)} & \text{non-singular} \\ & \uparrow & & \uparrow & \\ \longrightarrow \; . \; . \; . \; . \; . \; \longrightarrow & X_1 & \xrightarrow{h} & X^{(2)} & \end{array}$$

$x \in X$, an isolated ν-fold point[1] with $e = \dim \mathcal{T}_x(X) = 1$. We show by contradiction that

(■) <u>At some finite stage there are no ν-fold points lying above x.</u>

Let $R = \mathcal{O}_{Z,x}$, $M = \mathfrak{m}_{Z,x}$, $R/_{(f)R} = \mathcal{O}_{X,x}$. In $\hat{R} = k[[y,u,t]]$ we may assume $f = f_\nu + \tilde{f}$ where $\nu_m(\tilde{f}) > \nu$ and f_ν is a form of degree exactly ν in y and u. Denote $\varphi = In_M(f) = In_M(f_\nu)$. (In other words, y and u are a basis for the vector space T defining the strict tangent space $\mathcal{T}_x(X)$ (See Page (3)) and t extends this basis to a system of parameters of R).

If there is no ν-fold point lying above x we are done. Assume, then, that $x' \in X_1$ is such a point. <u>Then Idea A asserts that it is the unique point of the fibre at which $y/t = u/t = 0$.</u> In fact, since $T_x(D) = 0$ $(D = \{x\})$, <u>Idea A</u> $\Longrightarrow x \in P(\mathcal{T}_x(X)) \hookrightarrow h^{-1}(x)$. And in the blowing up of Z with center x (in which the situation is embedded) $g^{-1}(x) = \mathrm{Proj}\,(k[y,u,t])$ and $P(\mathcal{T}_x(X)) = \mathrm{Proj}(k[y,u,t]/_{(y,u)})$ = the origin in Spec $(k[y/_t, u/_t])$. Moreover if $R' = \mathcal{O}_{Z_1,x'}$, then $(y/_t, u/_t, t)$ is a regular system of parameters in $\hat{R}'$. (Let $M' = \max.(R')$).

<u>Fact:</u> X_1 is defined in Z_1 at x' by $(f')R'$, $f' = f/_{t^\nu}$ [2]

---

(1). We are assuming that the sequence of monoidal transformations corresponds to the procedure of steps I, II, III, and that we are starting at a <u>Step III</u>. In particular, we may assume h is the <u>quadratic</u> transformation with center x.

(2). This is a general fact about monoidal tranformations of regular local rings whose center is a regular prime ideal. See [1], chapter III, §2, P 216.

Thus $f' = f_{\nu}/_{t}\nu + \tilde{f}/_{t}\nu$, where $f_{\nu/t}\nu$ is a form in $y/_t$ and $u/_t$, of degree ν, and $\tilde{f}/_t\nu$ is a power series in the new parameters $y/_t$, $u/_t$, and t, divisible by t.(1)

Remark: Suppose we could conclude that:

$$* \begin{cases} \text{either (a)}\ \nu_{M'}(\tilde{f}')\ (= \text{ord}_{y/_t, u/_t, t}(\tilde{f}')) > \nu \\ \text{or}\quad (b)\ \nu_{M'}(\tilde{f}') < \nu \\ \text{or}\quad (c)\ \dim \mathcal{S}_{x'}(X') = 0 \end{cases} \qquad (\tilde{f}' = \tilde{f}/t^{\nu})$$

Then if (b) or (c), we are done, because (c) $\Longrightarrow$ no ν-fold points can lie above x' at any subsequent stage of the sequence (Idea A), and (b) $\Longrightarrow$ $\nu_{M'}(f') = \nu_{M'}(\tilde{f}') < \nu$, i.e. $\nu_{x'}(X') < \nu$. And if (a), then we are in a situation that is exactly the same as the original (i.e. $\nu_{x'}(X') = \nu$, $e = 1$, and $In_{M'}(f')$ is a form in y' and u' (new parameters are $y' = y/_t$, $u' = u/_t$, $t' = t$.) Suppose then, that (a) recurs infinitely many times, i.e., there is an infinite sequence of points $x^{(m)} \varepsilon X$, each lying over the preceding one, and such that the equation defining $X_{(m)}$ in $Z_{(m)}$ at $x^{(m)}$ is:

$$f^{(m)} = f_{\nu}(y/_{t}m,\ u/_{t}m) + \tilde{f}^{(m)}$$

but $f^{(m)} = f/_{t}\nu m$, and we may assume $\tilde{f}^{(m)}$ involves no denominators in t larger than $t^{(\nu m-m)}$. (See Footnote (1)). Hence, multiplying by $t^{\nu m}$, obtain:

$$f\ \varepsilon\ (y,u)^{\nu} + (t)^{m}R.$$

And since this is true $\forall$ m, $f\ \varepsilon\ (y,u)^{\nu}R$, so original x lies on a ν-fold curve defined by $y = u = 0$. Contradiction.

---

(1) The point is that any term of type $t^a y^i u^j/_t\nu$ $(a+i+j>\nu, a\geq 1)$ may be written in the form $t^{a'}(y/_t)^i(u/_t)^j$ where $a'=a-\nu+i+j$. Moreover every such term effectively involves a power of t no greater than $\nu-1$ in the denominator.

At this point, we notice that:

1) There is a strong intuitive appeal for the truth of (∗).

2) (∗) is not true.

The point is that if neither possibilities a) or b) hold, then something in $\tilde{f}'$ must contribute to the initial form of f', but everything in $\tilde{f}'$ involves t, so a priori one needs all three parameters to express the initial form (since both y' and u' are involved in f_ν/t^ν). Hence one would like to conclude $\mathcal{T}_{x'}(X') = 0$. The difficulty lies in the possibility that there exists a new parametrization such that the initial form does not involve all three new parameters. In the proof (which follows), the concept of "γ-preparation" is introduced to handle this possibility.

<u>Now with assumptions and notations as above, we prove the assertion (■)</u> (p. 15).

We have a regular system of parameters (y,u,t) of $R=\mathcal{O}_{Z,x}$ such that if $\bar{y}$, $\bar{u}$, $\bar{t}$ are their initial forms, then

$$T_x(Z) = \operatorname{Spec} Gr_x(Z) = \operatorname{Spec} k[\bar{y},\bar{u},\bar{t}] (k = \kappa(x)),$$

and $\mathcal{T}_x(X) = \operatorname{Spec} (k[\bar{y},\bar{u},\bar{t}]/(\bar{y},\bar{u}))$.

Note $\hat{R} = k[[y,u,t]]$. $R \hookrightarrow \hat{R}$. (f)R = ideal of X in Z at x.

Write

$$f = \sum_{i,j} f_{ij}(t) y^i u^j \text{ , where } f_{ij}(t) = \sum_{a=0}^{\infty} F_{ija} t^a, F_{ija} \in k.$$

Note that the initial form φ of f is $\varphi = \sum_{i+j=\nu} F_{ij0} \bar{y}^i \bar{u}^j$

Now define:

$$\Delta = \Delta(f;y,u,t) = \bigcup_{\substack{i,j \\ i+j<\nu}} \left\{ \frac{a}{\nu-(i+j)} \mid F_{ija} \neq 0 \right\} \subset \mathbb{Q}$$

(Think of the $\frac{a}{\nu-(i+j)}$ as a "weighted system of exponents").

Let $\gamma = \gamma_{y,u,t}(f)$ be the smallest element of Δ(1), and let $|\Delta|$ = the "hull" of $\Delta = \{q \in Q | q \geq \gamma\}$.

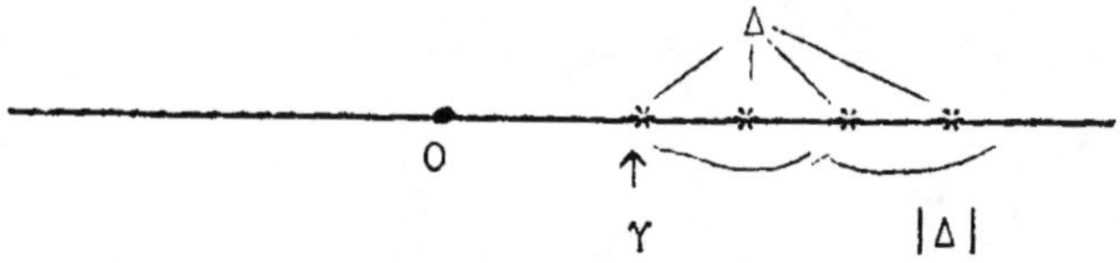

γ (and hence $|\Delta|$) is independent of the choice of the base f of (f)R, but it does depend on the choice of parameters.

<u>Remark:</u> γ, Δ are defined whenever we have expressions

(E): $f = \sum_{ij} f_{ij}(t)\, y^i u^j \ , \ f_{ij}(t) = \sum_a F_{ija} t^a$;

The important (and trivially verified) fact about γ is that given <u>any</u> such f,

$$\underline{\gamma_{y,u,t}(f) < 1 \iff \nu_M(f) < \nu}$$

Now let $S_\gamma = \{(i,j) / \frac{a}{\nu-(i+j)} = \gamma$ for some a such that $F_{ija} \neq 0\}$,

and <u>define</u>: $[f]_{y,u,t} = \varphi + \sum$ "terms contributing to γ", i.e.

$$= \varphi + \sum_{\substack{(i,j)\\ \in S_\gamma}} F_{ij\gamma(\nu-i-j)}\, \bar t^{\gamma(\nu-i-j)} \bar y^i \bar u^j$$

where $\varphi = In_M(f)$.

<u>Definition</u>: $[f]_{y,u,t}$ is solvable if $\exists\ \xi, \eta \in k$ such that

$$[f] = \varphi(\bar y - \xi \bar t^\gamma,\ \bar u - \eta \bar t^\gamma)$$

(Note that [f] solvable => γ is an integer).

If $[f]_{y,u,t}$ is solvable, we make a <u>γ-preparation</u>, i.e., we replace y and u by new parameters

$$y_1 = y - \xi t^\gamma,\ u_1 = u - \eta t^\gamma \ .$$

An easy computation then shows that:

$$\underline{\gamma_{y_1,u_1,t}(f) > \gamma_{y,u,t}(f)}$$

(1) The assumption that x is not on a ν-fold curve $y=u=0 \Rightarrow \Delta$ is non-empty and $\gamma>0$. In fact, there must be some term $F_{ija} t^a y^i u^j$ such that $F_{ija} \neq 0$ and $i+j<\nu$. And in such a term $a \geq 1$ (otherwise $\nu_m(f) < \nu$).

If $[f]_{y_1,u_1,t}$ is solvable, do the same. Suppose this is possible infinitely many times. We get:

$$(y_1,u_1)\ ,\ (y_2,u_2),\ldots,(y_m,u_m),\ldots$$

$$\gamma_1 < \gamma_2 < \ldots < \gamma_m < \ldots$$

Since each γ_i is an integer, the sequences y_i and u_i converge in the M-adic topology. Let $y^* = \lim y_i$, $u^* = \lim u_i$ (in $\hat{R}$). Then $f \in (y^*,u^*)^{\nu}\hat{R}$. (In fact, $\gamma_\infty = \gamma_{y^*,u^*,t}(f) = \infty$, so $F_{ija} = 0\ \forall i + j < \nu$). Thus x lies on a <u>formal ν-fold curve</u>. But this yields a contradiction via the following <u>General Fact</u>: Excellent $\Rightarrow$ ν-fold locus is compatible with completion (dimension remains the same).

<u>We may therefore assume F is γ-prepared</u> (i.e., not solvable).

Now recall our situation:

$$\begin{array}{ccc} Z_1 & \xrightarrow{g} & Z \\ \uparrow & & \uparrow \\ h:\ X_1 & \longrightarrow & X \end{array}$$

h = quadratic transformation with center x.

$$x' \longrightarrow x$$

both isolated ν-fold points

$\mathcal{T}_x(X) = \mathrm{Spec}(k[\bar{y},\bar{u},\bar{t}]/_{(\bar{y},\bar{u})})$. x' = the point $P(\mathcal{T}_x(X))$. In the fibre $h^{-1}(x)$, x' is the point $\bar{y}/_{\bar{t}} = \bar{u}/_{\bar{t}} = 0$, so $(y',u',t) = (y/_t, u/_t, t)$ is a regular system of parameters in $R' = \mathcal{O}_{Z_1,x'}$. The equation of X_1 in Z_1 at x' is

$$f' = f/_{t^\nu} = \sum f_{ij}(t)/_{t^{\nu-(i+j)}}(y/_t)^i(u/_t)^j$$

<u>The significant facts (verifiable by easy computations) are:</u>

(1). <u>$\gamma_{y',u',t}(f') = \gamma_{y,u,t}(f) - 1$</u>

(2). $\underline{[f']_{y',u',t} = [f]_{y,u,t}/\bar{t}^{\nu}}$

(Where $Gr_{x'}(Z_1) = k[\bar{y}',\bar{u}',\bar{t}]$, $Gr_x(Z) = k[\bar{y},\bar{u},\bar{t}]$,

and we identify $\bar{y}' = \bar{y}/\bar{t}$, $\bar{u}' = \bar{u}/\bar{t}$).

<u>In particular, $[f']_{y',u't}$ solvable => $[f]_{y,u,t}$ solvable.</u>

(3). <u>If $\gamma_{y',u't}(f') > 1$, then $In_{M'}(f') = \varphi'$</u>

<u>$= \varphi(y',u')$</u> $(= \varphi/\bar{t}^{\nu}$, $\varphi = In_M(f))$.

<u>In particular the situation remains the same in this case</u>

(i.e., $\dim \mathcal{T}_{x'}(X_1) = 1$, $\nu_{x'}(X_1) = \nu$).

By (3) and (1), after a finite number of quadratic transformations, $\gamma \leq 1$. So may assume in fact that $\gamma' = \gamma_{y',u',t}(f') \leq 1$.

If $\gamma' < 1$, $\nu_{x'}(X_1) < \nu$ contradiction.

Suppose then that $\gamma' = 1$. Then the initial form of f' is equal to $[f']_{y',u',t}$. (i.e., the order of every term contributing to γ' is precisely ν). Then there are two possibilities (keeping in mind that t, and hence the initial forms of <u>all three</u> parameters, appear non-trivially in $In_{M'}(f') = [f']_{y',u',t}$):

(i) $\mathcal{T}_{x'}(X_1) = 0$ (which is a contradiction because then, by <u>Idea A,</u> no more ν-fold points appear).

(ii) $\exists$ parameters r,s in R' such that $[f']_{y',u',t} = \psi(\bar{r},\bar{s})$, where ψ is a form of degree ν. ($\bar{r} = In_{M'}(r)$, $\bar{s} = In_{M'}(s)$)

But (ii) yields a contradiction as follows:

we may assume $\bar{r} = (a\bar{y}' + b\bar{u}' + c\bar{t})$, $\bar{s} = (a'\bar{y}' + b'\bar{u}' + c'\bar{t})$.

Have $\varphi(\bar{y},\bar{u}) + \bar{t}\cdot(\text{something}) = [f]_{y,u,t} = [f']_{y',u',t}\cdot \bar{t}^{\nu}$

$$= \psi(\bar{r},\bar{s})\cdot\bar{t}^{\nu} = \psi(\bar{r}\bar{t},\bar{s}\bar{t}) = \psi((a\bar{y}+b\bar{u}+c\bar{t}^2),(a'\bar{y}+b'\bar{u}+c'\bar{t}^2))$$

Therefore $\varphi(\bar{y},\bar{u}) = \psi((a\bar{y}+b\bar{u}),(a'\bar{y}+b'\bar{u}))$.

Notice then that $ab'-ba' \neq 0$, for otherwise $\varphi(\bar{y},\bar{u})$ is a form of degree ν in one parameter $a\bar{y} + b\bar{y}$ ($=d(a'\bar{y}+b'\bar{u})$, some $d \in k$) and contradicting $e = 1$. Now use

<u>Lemma</u>: Let $\varphi = \varphi(X,Y), \psi = \psi(Z,W)$ be forms of degree ν in two variables over k, and suppose that $\varphi(y,u) = \psi((ay+bu),(a'y+b'u)) \in k[y,u]$, where $a,a',b,b' \in k$ and $ab' - ba' \neq 0$. Then $\forall\, c,c' \in k\ \exists\ \xi, \eta \in k$ such that, (in $k[y,u,t]$) $\varphi(y+\xi t, u+\eta t) = \psi((ay+bu+ct),(a'y+b'u+c't))$.

<u>Proof</u>: $\varphi(y+\xi t,u+\eta t) = \psi(a(y+\xi t)+b(u+\eta t), a'(y+\xi t)+b'(u+\eta t))$

$$= (ay+bu+(a\xi+b\eta)t, a'y+b'u+(a'\xi+b'\eta)t)$$

So we want to solve:

$$\left.\begin{aligned} a\xi + b\eta &= c \\ a'\xi + b'\eta &= c' \end{aligned}\right\} \text{ for } \xi \text{ and } \eta$$

And this can be done if $ab'-ba' \neq 0$.

QED

Thus (letting $t = t^2$ in lemma and recalling that $\gamma' = 1 \iff \gamma = 2$) (ii) $\Longrightarrow$ <u>$[f']$ solvable $\Rightarrow [f]_{y,u,t}$ is solvable</u>, which is a contradiction.

<u>This completes the proof of (■) for the case e = 1.</u>

2. <u>e = 2</u> (cusp situation)

As before, we suppose that we have an infinite sequence $\{h_{(m)}\}$ of permissible monoidal transformations:

$$\begin{array}{ccccccccc}
\cdots \xrightarrow{g(m)} & Z_{(m)} & \longrightarrow & \cdots\cdots & \longrightarrow & Z_1 & \xrightarrow{g} & Z & \\
 & \uparrow & & & & \uparrow & & \uparrow & \\
\cdots \xrightarrow{h(m)} & X_{(m)} & \longrightarrow & \cdots\cdots & \longrightarrow & X_1 & \xrightarrow{h} & X & \\
\cdots \longrightarrow & x^{(m)} & \longrightarrow & \cdots\cdots & \longrightarrow & x' & \longrightarrow & x & \text{all } \nu\text{-fold points}
\end{array}$$

where now <u>x is an isolated ν-fold point such that $\dim \mathcal{T}_x(X)=2$.</u>

$R = \mathcal{O}_{Z,x}$, $k = \kappa(x)$, f defines X in Z at x. $M = \max(R)$

Choose a system of parameters (y,u,t) of R so that

$$\mathcal{T}_x(X) = \mathrm{Spec}(k[\bar{y},\bar{u},\bar{t}]/_{(\bar{y})}) \quad (\bar{y} = \mathrm{In}_M(y)) \text{ etc.}$$

Corresponding to the data $(f;y,u,t)$, write

(F): $$f = \sum_i g_i(t,u)y^i$$

$$g_i(u,t) = \sum_{c,d} G_{icd}t^c u^d \quad (G_{icd} \in k)$$

(Note: $\mathrm{In}_M(f) = \varphi = G_{\nu,o,o}\bar{y}^{\nu}$)

Define

$$\Delta = \Delta(f;y,u,t) = \bigcup_{i=o}^{\nu-1} \left\{ \begin{matrix}(a,b)\\ \in Q^2\end{matrix} \mid G_{i,a(\nu-i)b(\nu-i)} \neq 0 \right\}$$

$$= \bigcup_{i=o}^{\nu-1} \left\{ \frac{c}{\nu-i}, \frac{d}{\nu-i} \mid G_{i,c,d} \neq 0 \right\}$$

and let

$|\Delta|$ = smallest convex set in $\mathbb{R}^2$ such that $\Delta \subset |\Delta|$

and $(a,b) \in |\Delta| \Longrightarrow (a+r,b+s) \in |\Delta| \; \forall \; r,s \geq 0$.

Let $r \in \mathbb{R}$. We will use the notation:

$S(r)$ = line through $(r,0)$ with slope -1.

$V(r)$ = vertical line through $(r,0)$.

<u>Remark</u>: (1) $\Delta = \Delta(f;\, y,u,t)$ is defined whenever we have expressions (F) (i.e., without reference to any special kind of $x \in X \hookrightarrow Z$). Then it is easy to check that

(i) The vertices of $|\Delta|$ are points of Δ, and they all lie on the <u>lattice</u>

$$\mathbb{Z}(1/_n) \times \mathbb{Z}(1/_n) \hookrightarrow \mathbb{R}^2 \quad (n = \nu!)$$

(ii) <u>$\nu_M(f) < \nu \Longleftrightarrow |\Delta|$ contains a point on $S(r)$ with $r < 1 \Longleftrightarrow$ there is a vertex (a,b) with $a + b < 1$</u>.

(iii) Let <u>$\alpha = \alpha_{y,u,t}(f)$</u> = the distance of $|\Delta|$ from the b-axis = the smallest "a" appearing in any $(a,b) \in |\Delta|$. Then $\alpha < 1 \Longleftrightarrow y = t = 0$ is not a ν-fold curve through the point defined by f.

(iv) Some vertex of $|\Delta|$ lies below the line

$$b = 1 \iff y = u = 0 \text{ is not a } \nu\text{-fold curve.}$$

With $\alpha = \alpha_{y,u,t}(f)$ as in Remark (iii) let $\beta = \beta_{y,u,t}(f)$ be the smallest such that $(\alpha,\beta) \in |\Delta|$. ($(\alpha,\beta)$ is then necessarily a vertex). Let $\gamma = \gamma_{y,u,t}(f)$ be the smallest number such that $S(\gamma) \cap |\Delta| \neq \emptyset$, and let $\delta = \delta_{y,u,t}(f)$ be such that $(\gamma-\delta,\delta)$ is the lowest point on $S(\gamma) \cap |\Delta|$. (α,β) and $(\gamma-\delta,\delta)$ are the "most important" vertices.

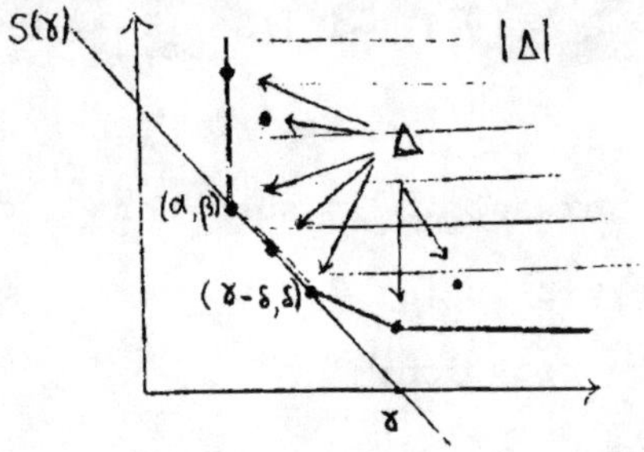

Solvability, (a,b) - preparations and Well-Preparedness

Let (a,b) be any vertex of $|\Delta|$, and define

$$\{f\}^{a,b}_{y,u,t} = \varphi + \sum \text{"terms contributing to (a,b)"}$$

$$= \sum_{0 \le i \le \nu} G_{i,a(\nu-i),b(\nu-i)}\bar{y}^{i}\bar{t}^{a(\nu-i)}\bar{u}^{b(\nu-i)}$$

We say $\{f\}^{a,b}_{y,u,t}$ is solvable (or equivalently that the vertex (a,b) is unprepared) if a,b are integers and

$$\{f\}^{a,b}_{y,u,t} = \varphi(\bar{y} - \xi\bar{t}^{a}\bar{u}^{b})$$

For some $\xi \in k$. If $\{f\}^{a,b}_{y,u,t}$ is solvable, we can make an "(a,b)-preparation" which is by definition a change of parameter

$$y \longrightarrow y_1 = y - \xi\, t^a u^b$$

If we cannot make an (a,b)-preparation for any vertex

(a,b) of $|\Delta|$ (i.e., if all the $\{f\}^{a,b}_{y,u,t}$ are unsolvable) then say we are "Well Prepared" ("W.P.")

Remark: (2) If $y \longrightarrow y_1$ is an (a,b)-preparation, then

(1) $|\Delta(f;y_1,u,t)| \subset |\Delta(f;y,u,t)| - \{(a,b)\}$.

(The vertex (a,b) is eliminated).

(2) If $(a',b') \neq (a,b)$ is another vertex, $y \longrightarrow y_1$ leaves it intact. (i.e., (a',b') is a vertex of $|\Delta(f;y_1,u,t)|$, and $\{f\}^{a',b'}_{y_1,u,t} = \{f\}^{a',b'}_{y,u,t}$ in the sense that the former can be obtained by substituting the symbol $\bar{y}_1$ for the symbol $\bar{y}$ in the latter).

Note: by the above, and the fact that $|\Delta|$ has only a finite number of vertices, and by an argument analogous to that of Page (19), we deduce:

By a finite number of vertex preparations we may always obtain a W.P. situation.

Translations in u and Very-Well Preparedness

In addition to the vertex preparations described above, we will also need to consider translations of the type:

$$u \longrightarrow u_1 = u - \xi t^n, \quad \xi \in k, \ n \text{ a positive integer.}$$

Remark: (3) One checks easily that under $u \longrightarrow u_1$, a term in f corresponding to a point $(a,b) \in |\Delta|$ can affect only those points strictly below it on the line of slope $-1/n$ through it. In particular

$$\alpha_{y,u,t}(f) = \alpha_{y,u_1,t}(f)$$

$$\beta_{y,u,t}(f) = \beta_{y,u_1,t}(f)$$

and $\{f\}^{\alpha,\beta}_{y,u,t}$ is left intact.

Now suppose that $(f;y,u,t)$ is W.P. (can always assume to be starting with this situation). Then choose $\xi \in k$ so that after $u \longrightarrow u_1 = u-\xi t$ (and subsequent W-preparation $y \longrightarrow y_1$) we obtain the <u>largest possible δ</u>.

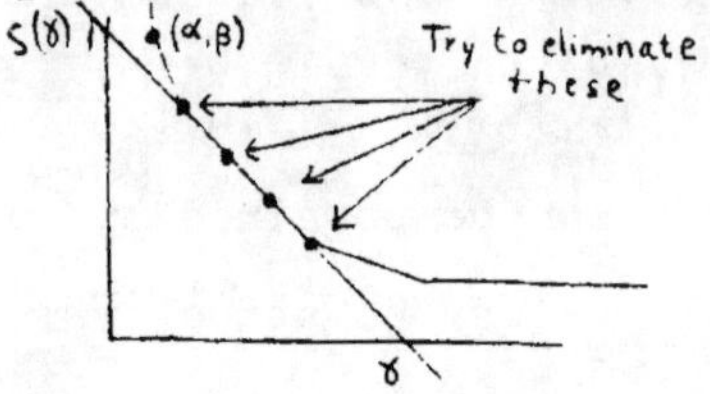

Observe that by the previous <u>Remark</u> the highest point on $S(\gamma)$ must remain, so that, if y_1, u_1 are the new parameters,

$$\gamma_{y_1,u_1,t}(f) = \gamma_{y,u,t}(f).$$

Having obtained the largest possible δ in the above sense, we distinguish two cases:

<u>Case I</u> <u>$(\gamma-\delta,\delta) \neq (\alpha,\beta)$</u>. Then we say that $(f;y,u,t)$ is "<u>Very-Well-Prepared</u>" ("V.W.P.")

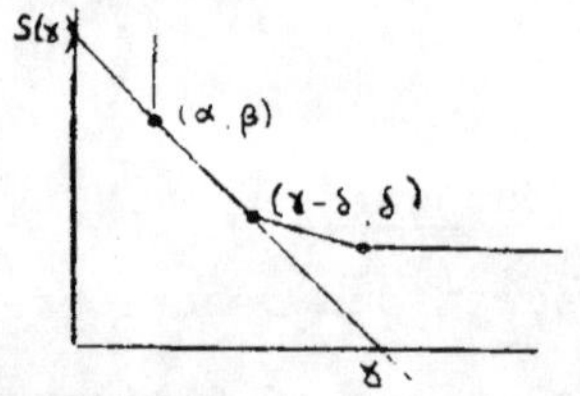

<u>Case II</u> <u>$(\gamma-\delta,\delta) = (\alpha,\beta)$</u>. Then consider the line $U_{\alpha\beta}$ through (α,β) with the largest slope such that none of $|\Delta|$ lies below it. Let $\mathcal{E} = |$ slope of $U_{\alpha\beta}|$. (Note that $\mathcal{E} < 1$), and let $(\varkappa',\varkappa)$ be the lowest point on $U_{\alpha\beta} \cap |\Delta|$.

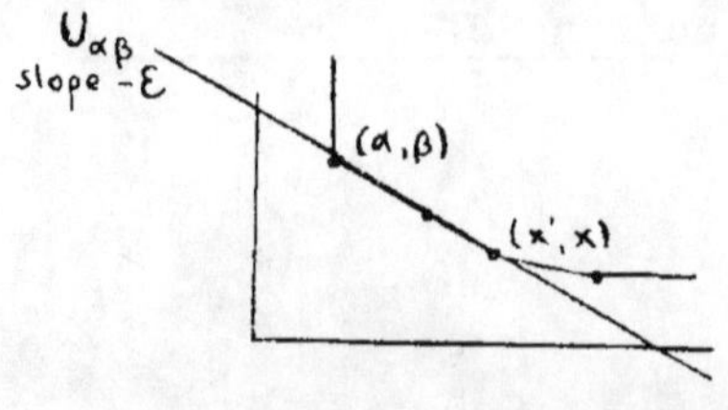

If $1/\varepsilon$ is not an integer, say we are V.W.P.

If $1/\varepsilon$ is an integer, choose $\xi \in k$ such that after $u \longrightarrow u_1 = u - \xi t^{1/\varepsilon}$ and subsequent well-preparation. The number $\varkappa$ becomes as large as possible (see Remark(3), Page 26)). If we get to $\varkappa = \beta$, get new $U_{\alpha\beta}$ and repeat. We can undergo this procedure only finitely many times, and when we obtain the largest possible $\varkappa$, say we are V.W.P.

<u>Effects of Monoidal Transformations</u>; <u>Idea B</u>: <u>The Sequence of Monoidal Transformations "Sharpens" $|\Delta|$</u>.

Given the date $(f;y,u,t)$ we have the following operations:

(i) <u>Vertex Preparations at (a,b)</u> $(y \longrightarrow y_1 = y - \xi t^a u^b)$.
We have remarked that this leaves other vertices (a',b') intact, along with their data $\{f\}^{a'b'}_{y,u,t}$.
<u>Translations in u</u> $(u \longrightarrow u_1 = u - \xi t^{1/\varepsilon})$
We have remarked that under this operation a point (a,b) contributes only to points strictly below it on the line of slope $-\varepsilon$ through it.

(ii) <u>Quadratic and monoidal transformations</u>.

<u>Remark (4)</u> If we assume $(f;y,u,t)$ is (α,β) - prepared, then β is not affected by any operation in (i) above. In particular β is not affected by passing from a W. P. to a V. W. P. situation. We propose to prove <u>Idea B</u> in the form: <u>By the sequence $\{h_{(m)}\}$ of monoidal transformations β is non-increasing and eventually decreases</u>. (This implies in essence that we eventually arrive at $\alpha + \beta < 1$, so (by Remark 1)(ii), Page (24)) the multiplicity drops.)

Consider a permissible monoidal transformation

$$\begin{array}{ccc} Z' & \xrightarrow{g} & Z \\ \uparrow & & \uparrow \\ X' & \xrightarrow{h} & X \\ \\ x' & \longrightarrow & x \end{array}$$

x, x' are ν-fold points
dim $\mathcal{T}_x(X) = 2$.

(Think of this as occuring at some arbitrarily selected stage in our sequence $\{h_{(m)}\}$; in particular x need not be isolated in the ν-fold locus).

Let $R, M = \mathcal{O}_{Z,x}$ $\quad R', M' = \mathcal{O}_{Z',x'}$

(y,u,t) and (y',u',t') are a system of parameters for R and R' respectively.

$(f;y,u,t)$ and $(f';y',u',t')$ are data of X in Z at x and X' in Z' at x' respectively.

We assume that $\mathcal{T}_x(X) = \mathrm{Spec}(k[\bar{y},\bar{u},\bar{t}]/(\bar{y}))$ and that $(f;y,u,t)$ is V.W.P. (As usual, $\bar{y}$ denotes $\mathrm{In}_M(y)$, etc.)

By Idea A there are four relevant ways in which x' can be related to x:

T-1 h is a quadratic transformation with center x and $(y',u',t') = (y/_t, u/_t - \zeta, t)$, $\zeta \in k$.

T-2 h is a quadratic with center x and $(y',u',t') = (y/_u, u, t/_u)$

T-3 h is a monoidal transformation with cent. $y = t = 0$ and $(y',u',t') = (y/_t, u, t)$

T-4 h is monoidal with center $y = u = 0$ and $(y',u',t') = (y/_u, u, t)$

Since y defines $\mathcal{T}_x(X)$, and since we are assuming that we have reached a stage in the resolving sequence where the only irreducible ν-fold curves are non-singular (see Page (12)) we may always choose u and t so that one of the above occurs.

We now investigate the effect of each of the above on β.

We will denote:

$(\alpha,\beta) = (\alpha_{y,u,t}(f), \beta_{y,u,t}(f))$ $(\alpha',\beta')=(\alpha_{y',u',t'}(f'),\beta_{y',u',t'}(f'))$

$\sigma(\Delta(f;y,u,t)) = \Delta(f';y',u',t')$, a transformation in the plane.

T-3 Here $f'=f/_t\nu$, and σ is a translation to the left by 1 in fact:

$$y^i t^c u^d \longrightarrow (y/_t)^i u^d t^{c-(\nu-i)} = (y')^i (u')^d (t')^{c-(\nu-i)}$$

so the point $(a,b) = (\frac{c}{\nu-i}, \frac{d}{\nu-i})$ moves to the point $(\frac{c-(\nu-i)}{\nu-i}, \frac{d}{\nu-i}) = (a-1,b)$.

In particular $\sigma(\alpha,\beta) = (\alpha',\beta')$ (since σ preserves the shape and the orientation of $|\Delta|$). $\therefore (\alpha',\beta') = (\alpha-1,\beta)$: β does not change.

Note that T-3 occurs only when $\alpha \geq 1$. Conversely, at any point x whose α is ≥ 1, the only kind of monoidal transformation in the sequence $\{h_{(m)}\}$ which affects it is T-3 (because of the construction of the sequence $\{h_{(m)}\}$; see Page (14)). Hence there can be only a finite number of successive T-3's.

T-4. See similarly that σ is a translation down by 1; $\sigma(\alpha,\beta) = (\alpha',\beta')$ so $\beta' = \beta-1$: β decreases.

Note that T-4 occurs only when none of $|\Delta|$ lies below the line $b = 1$; conversely the only $h_{(m)}$'s affecting x's whose $|\Delta|$ have this property are T-4; there can be only a finite number of successive T-4's.

T-2. One checks that $\sigma(a,b) = (a,b+a-1)$. Therefore points move vertically and the lines $S_{(r)}$ are transformed into horizontal lines $b = r-1$. (In general lines of negative slope are transformed into lines with increased slope). We again have $\sigma(\alpha,\beta) = (\alpha',\beta')$. so $(\alpha',\beta') = (\alpha,\beta+\alpha-1)$. But

$\alpha < 1$ (otherwise T-3) so $\beta' < \beta$: β decreases.

Remark (5). In all the above cases, if $(a,b) \in \Delta$, and $\sigma(a,b) = (a',b')$, then

$$\{f'\}^{a'b'}_{y',u',t'} = \begin{cases} \{f\}^{ab}_{y,u,t}/\bar{t}^{\nu} & \text{T-3} \\ \{f\}^{ab}_{y,u,t}/\bar{u}^{\nu} & \text{T-2, T-4} \end{cases}$$

Thus $(f;y,u,t)$ is (a,b)-prepared $\Rightarrow (f';y',u',t')$ is (a',b')-prepared.

In particular W.P. is preserved. Note that from a W.P. situation we can pass to a V.W.P. situation without affecting (α,β).

T-1 (Main Case).

a) $\zeta \neq 0$ (Here something good happens because we may assume to be V.W.P.). Apply the translation $u \longrightarrow u_1 = u - \zeta t$ and well-prepare (via some $y \longrightarrow y_1$).

Note that this does not affect (α,β) but it might change $(\gamma-\delta,\delta)$ (See Remark (3) Page (26). Denote

$$\delta_1 = \delta_{y_1,u_1,t}(f) \qquad \gamma_1 = \gamma_{y_1,u_1,t}(f) .$$

Suppose we were originally (i.e., before $y \longrightarrow y_1, u \longrightarrow u_1$) in the situation of

Case I (Page (27) $(\alpha,\beta) \neq (\gamma-\delta,\delta)$. In particular have $\delta < \beta$. And since δ is defined for a V.W.P. situation, $\delta_1 \leq \delta < \beta$.

Suppose we were originally in the situation of

Case II $(\alpha,\beta) = (\gamma-\delta,\delta)$. Then one can check that $(\gamma,0)$ is a vertex of $\Delta(f;y,u_1,t)$ and it is prepared[1] (and hence is

(1) A sketch of the proof of this assertion is: $(y/_t, u/_t-\zeta, t)$ is a ν-fold point $\Rightarrow$ every term in y, u_1, and t in the power series $\sum G_{icd}\, y^i t^c (u_1+\zeta t)^d$ (obtained by the substitution $u=u_1+\zeta t$ in the original expression of f) has order $\geq \nu$. In particular there are terms like $y^i t^{c+d}$ coming from those $y^i t^c u^d$ which contributed to (α,β). Such $y^i t^{c+d}$ are precisely those contributing to $(\gamma,0)$ in $\Delta(f;y,u_1,t)$. Then $\{f\}^{\gamma,0}_{y,u_1,t} \neq 0$ and $\{f\}^{\gamma,0}_{y,u_1,t}$ unsolvable follow essentially because it is obtained from $\{f\}^{\alpha\beta}_{y,u,t}$ by substituting ζt for u.

not affected by the subsequent $y \longrightarrow y_1$). Then we must have $(\gamma_1-\delta_1,\delta_1) = (\gamma,0)$, so $\delta_1 = 0 < \beta$ ($0<\beta$ because if $\beta = 0$, since $\alpha < 1$ must have $\alpha + \beta < 1$, so multiplicity drops - contradiction).

The point is that we have reduced T-1 $\zeta \neq 0$ to the case T-1 $\zeta = 0$, Case I $(\alpha,\beta) \neq (\gamma-\delta,\delta)$ (which will be treated next), except that we may no longer be V.W.P. This is inessential however - we are still W.P. (the reduction to $\zeta = 0$ is $u \longrightarrow u_1$).

T-1 b) $\zeta = 0$

Here $\sigma(a,b) = (a+b-1,b)$; points move horizontally, lines $S_{(r)}$ become $V_{(r-1)}$ and in general lines ofnegative slope decrease their slope. (See picture). Again we have, that if $\sigma(a,b) = a',b')$, then $\{f\}^{a'b'}_{y',u',t'}=\{f\}^{a,b}_{y,u,t/t^{\nu}}$, so W.P. is preserved, and hence so is V.W.P., since $(\gamma'-\delta',\delta)$ must be some $(\varkappa',\varkappa)$. (See Page (לז).

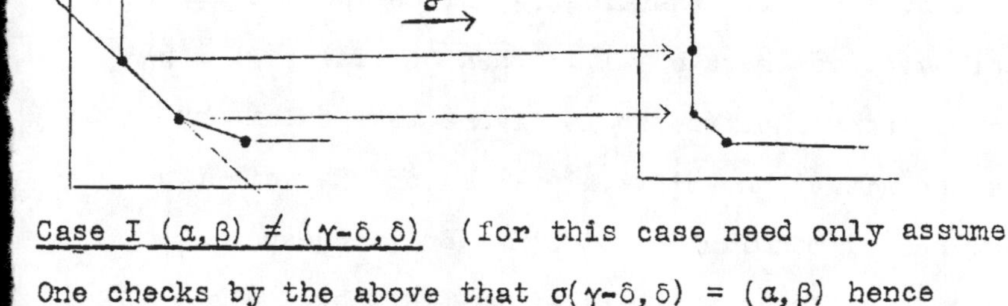

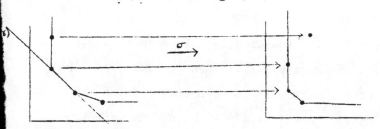

Case I $(\alpha,\beta) \neq (\gamma-\delta,\delta)$ (for this case need only assume W.P.). One checks by the above that $\sigma(\gamma-\delta,\delta) = (\alpha,\beta)$ hence $\beta' = \gamma < \beta$: β decreases.

Case II $(\alpha,\beta) = \gamma-\delta,\delta)$.

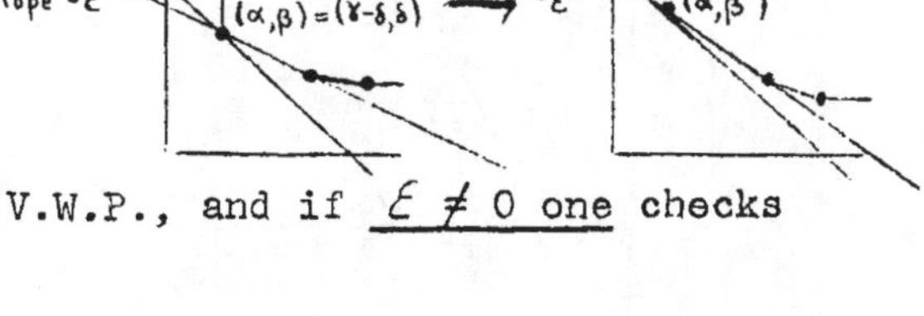

$(f';y',u',t')$ is again V.W.P., and if $\varepsilon \neq 0$ one checks easily that $1/\mathcal{E}' = 1/\mathcal{E} - 1$. (where $\mathcal{E}'$ is the absolute value of the slope of the line through $(\alpha',\beta') = (\alpha,\beta)$ of largest slope having none of $|\Delta'|$ below it). In particular $\mathcal{E}' > \mathcal{E}$, and eventually $\mathcal{E}' \geq 1$; then we will be in T-1, $\zeta = 0$, Case I.

If the sequence of monoidal transformations involves only the situations treated above, we will have an infinite number of decreases of β. But since (α,β) moves on a lattice, we must eventually reach $\beta = 0$, after which there can be no further decreases - contradiction.

Thus we need only be concerned with the occurence of the one remaining situation:

T-1, $\zeta = 0$, $(\alpha,\beta) = (\gamma-\delta,\delta)$ and $\mathcal{E} = 0$.

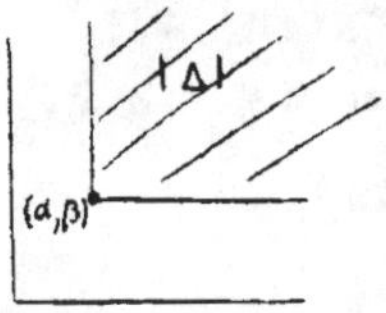

Here we must have $\beta < 1$ (otherwise T-4 since in this case $\beta \geq 1 \Rightarrow$ all vertices have $b \geq 1$). Then $(\alpha',\beta') = (\alpha+\beta-1,\beta)$ and in fact σ translates the whole figure to the left by $1-\beta$. Thus the situation $(f';y',u',t')$ is of the same type, so the sequence of monoidal transformations corresponds to a sequence of translations of $|\Delta|$ to the left by $1 - \beta$. Therefore (since $\alpha < 1$ otherwise T-3) we eventually come to $\alpha + \beta \leq 1$, i.e., we get into the shaded area:

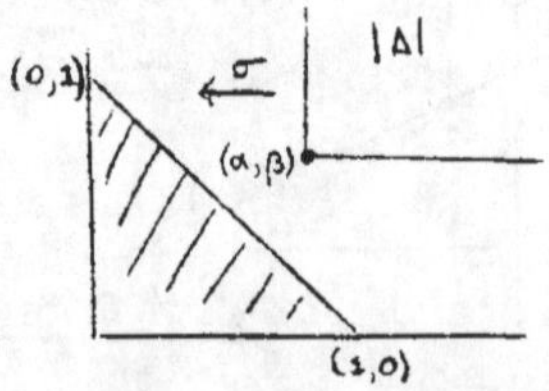

If we actually get $\alpha + \beta < 1$, then the multiplicity drops - contradiction. If we get a boundary point (the only one to worry about is really (0,1)), then by an argument analogous to that for the case $e = 1$ (Page (21)), we obtain $\dim \mathcal{T}_x(X) < 2$, so we are reduced to the case $e = 1$.

Q.E.D.

Remarks about the general theorem

Starting with a two dimensional excellent X (without any hypothesis about $X \hookrightarrow Z$) the structure of the induction essentially allows us to reduce to the analysis of the behavior of an isolated singular point under successive monoidal transformations. Thus we may consider just a two dimensional excellent local ring with only isolated singularity (i.e., $\mathcal{O}$ localized at any non maximal prime is regular), and, by the general theory of monoidal transformations, we may even assume $\mathcal{O}$ is complete. Then we have an embedding $X = \mathrm{Spec}(\mathcal{O}) \longrightarrow \mathrm{Spec}(R)$ where (R,M) is a complete regular local ring. Thus $\mathcal{O} = R/_J$, where now $J = (f_1, \ldots, f_n)$.

1) Instead of $\nu_x(X)$ we now use $\nu_x^*(X)$ or equivalently $\nu_x^*(X,Z)$, which is the sequence:

$(\nu_M(f_1), \ldots, \nu_M(f_r), \infty, \infty, \ldots)$

(assuming the generators f_i of J are suitably chosen. In particular, they must be a "standard base").

2) $|\Delta(J;y,u,t)|$ may be defined in this more general setting. (The y,u,t now stand for tuples of parameters).

3) Idea A works (even with imperfect residue field extension). In general the statement "The singularity of x' in X' is as bad as the singularity of x in X must be taken to mean:

$$\nu^*_{x'}(X') \geq \nu^*_x(X) \quad \underline{\text{lexicographically}} \quad .$$

APPENDIX

We give a sketch of a proof of Idea A (in the special case we have treated):

(R,M) a regular, local k-algebra. $R/_M = k$, $k = \bar{k}$, $P \hookrightarrow R$ a regular prime ideal, $f \in R$. Suppose that $\nu = \nu_M(f) = \nu_P(f)$. (This is normal flatness). Let (R',M') be a monoidal transform of R with center P such that $PR' = tR'$(1) and $R'/_{M'} = R/_M = k$. Suppose that $\nu_{M'}(f/_t\nu) \geq \nu$. Then if $(y_1,\ldots,y_r)$ are part of a system of parameters of R such that

$$T = k\mathrm{In}_M(y_1) + \ldots + k\mathrm{In}_M(y_r)$$

is the smallest subspace of Gr^1_M for which $\mathrm{In}_M(f) \in k[T]$, we have:

$$\underline{y_1/_t,\ldots,y_r/_t \text{ are regular parameters in } R'} \quad .$$

Proof. By Normal Flatness we may assume the y_i are in P. (In fact, by condition (2) of the Theorem on Normal Flatness, page (4), we have "$T_x(D) \hookrightarrow \mathfrak{T}_x(X)$", so the subspace T of $Gr^1_M(R)$ defining $\mathfrak{T}_x(X)$ is contained in the subspace defining $T_x(D)$. But in our case $D = V(P) \hookrightarrow \mathrm{Spec}\ R$, so the subspace

(1) By definition, a monoidal transform of R with center P is a local R-algebra R' which is a localization of a ring of type $A = [p_1/_t,\ldots,p_n/_t]$ with respect to a prime ideal N containing MA. (where the p_i are a set of generators of P, $t \in P$, and A is regarded as an R-subalgebra of R_t). Then one can show that R' is necessarily the local ring of some point lying over the closed point in the monoidal transformation of Spec(R) with center V(P). (This assertion means we may assume $t = p_i$ for some i).

defining $T_x(D)$ is spanned by the initial forms of the parameters which generate P). So we have:

$P = (y_1,\ldots,y_r,x_1,\ldots,x_s)R$, part of a regular system of parameters of R. (And we may assume that t is an x or a y).

Now by the assumption $\nu = \nu_M(f) = \nu_P(f)$, we may write:

$$f = \sum_{\substack{\sigma=(\sigma_1,\ldots,\sigma_r)\\ |\sigma|=\nu}} a_\sigma y_1^{\sigma_1}\ldots y_r^{\sigma_r} + p \quad \text{where } a_\sigma \in k \text{ and } p \in P^{\nu+1}$$

so $$f/_t\nu = \sum_\sigma a_\sigma (y_1/_t)^{\sigma_1}\ldots(y_r/_t)^{\sigma_r} + p/_t\nu$$

<u>Claim that $\nu_{M'}(f/_t\nu) \geq \nu$ implies that all the $y_i/_t$ are in M'.</u>

(Then, by the general theory of monoidal transformations, it follows that the $y_i/_t$ are in fact <u>regular parameters</u> of R', which completes the proof). We know that for some $\xi = (\xi_1,\ldots,\xi_r) \in k^r$, $Z_i = y_i/_t - \xi_i$ are regular parameters of R'. We wish to show all the $\xi_i = 0$. If not, let $y/_t = (y_1/_t,\ldots,y_r/_t)$, $Z = (Z_1,\ldots,Z_r)$ and let F(Y) be the form

$$\sum_\sigma a_\sigma Y_1^{\sigma_1}\ldots Y_r^{\sigma_r}, \text{ (of degree } \nu \text{ over } k).$$

Since $\nu_{M'}(f/_t\nu) \geq \nu$, $F(y/_t) = F(Z+\xi) \in M'$. But this can only happen if $F(Z+\xi) = F(Z)$ (i.e., if after expanding $F(Z+\xi)$ all terms involving Z with total order $< \nu$ are 0). But then, by an easy argument about forms, after a suitable change of coordinates we can express F as a form in fewer than r variables, which leads to a contradiction regarding the dimension of T.

Q.E.D.

Bibliography

[1] Heisuke Hironaka, "Resolution of Singularities of an Algebraic Variety over a field of characteristic 0". Annals of Mathematics, Vol. 79, No.2, March 1964.

[2] ________ "On the Characters ν^* and t^* of Singularities". Columbia University Mimeographed Notes.